Inorganic Chemistry Through Experiment

BY

G. F. LIPTROT, M.A., Ph.D.
Head of the Chemistry Department, Eton College

MILLS & BOON LTD
17–19 FOLEY STREET
LONDON W1A 1DR

First published in Great Britain 1974 by
Mills & Boon Ltd, 17–19 Foley Street, London W1A 1DR.

© G. F. Liptrot 1975

ISBN 0 263 51591.5

All rights reserved. No part of this publication may be reproduced stored in a retrieval system, or transmitted in any form or by any means, electronic, mechanical, photocopying or recording or otherwise, without the prior permission of Mills & Boon Limited.

Made and printed in Great Britain by Thomson Litho Ltd., East Kilbride

Preface

Over the past twenty or so years there has been a steady advance in the methods of presentation of inorganic chemistry and a shift of emphasis has taken place, away from the traditional preparation and analysis of compounds for their own sake, towards a search for underlying patterns of behaviour. The author hopes that this book may indicate how practical inorganic chemistry can be taught within the broad framework of the Periodic Table and how additional physiochemical data, such as redox potentials, might profitably be employed to correlate chemical behaviour.

The emphasis has been placed on the use of rapid test-tube reactions and the conventional volumetric analysis, which employs burettes and pipettes, has been replaced by a dropping tube technique, as and when it has been felt desirable to illustrate the stoichiometric nature of some important reaction in solution. While the bulk of the experiments can be performed on a small scale, thus economising on chemicals, the conventional larger scale preparative work has not been neglected altogether; in particular, a number of preparations illustrating transition metal chemistry has been included. In addition, some experiments illustrating various unrelated aspects of an element's behaviour have been felt worthy of attention.

It is hoped that the book will prove useful to pupils studying both the new and the more traditional syllabuses at A-level and National Certificate level; to this end, a blend of experiments ranging from those of an 'open-ended' nature to those of a more didactic kind have been included, and it is hoped that a reasonable balance has been achieved.

Eton College 1974   G. F. LIPTROT

Acknowledgements

This book is the result of many years' experience in teaching inorganic chemistry at sixth form level, and the experiments have been accumulated over a long period of time; consequently the origin of many of these has now become rather obscure. However, I have been influenced considerably by recent changes in the shift of emphasis in inorganic chemistry and I should like to pay tribute to the Nuffield Advanced Chemistry Course which has proved to be a rich source of ideas, many of which have filtered through into this book. I am also grateful to Dr. D. J. Waddington for providing me with the details of the experiments on cobalt(III) complexes in Section 11.5, pages 170–173.

Finally I should like to thank Mr. D. Hughes for providing the three plates and for allowing me to draw on his many years of practical experience, Dr. D. J. Waddington and Mr. G. W. Walker for reading the whole of the book in proof and for being constructive in their criticism of certain parts, my wife who as usual undertook the trying task of typing the original manuscript, and my publishers for help at all stages in the production of this book.

Contents

Acknowledgements	4
Preface	3

Chapter

1.	The Periodic Classification of the Elements	7
2.	Group 1A The Alkali Metals	20
3.	Group 2A The Alkaline Earth Metals	30
4.	Group 3B Boron, Aluminium, Gallium, Indium and Thallium	39
5.	Group 4B Carbon, Silicon, Germanium, Tin and Lead	51
6.	Group 5B Nitrogen, Phosphorus, Arsenic, Antimony and Bismuth	73
7.	Oxidation-Reduction—Some Redox Reactions	96
8.	Group 6B Oxygen, Sulphur, Selenium, Tellurium and Polonium	106
9.	Group 7B Fluorine, Chlorine, Bromine, Iodine and Astatine	132
10.	The First Transition Series (Scandium, Titanium, Vanadium, Chromium, Manganese, Iron, Cobalt and Nickel)	148
11.	The First Transition Series in More Detail	162
12.	Group 1B Copper, Silver and Gold	178
13.	Group 2B Zinc, Cadmium and Mercury	190
	Appendix I: List of Chemicals	200
	Appendix II: Solutions Required (I)	203
	Appendix III: Solutions Required (II)	205
	Index	

List of Plates

1. Photographs of mica and a silicate rock containing asbestos.	60
2. The apparatus used to demonstrate the paramagnetism of manganese(II) sulphate	161
3. The apparatus used for colorimetric experiments	176

(Plates by courtesy of D. Hughes)

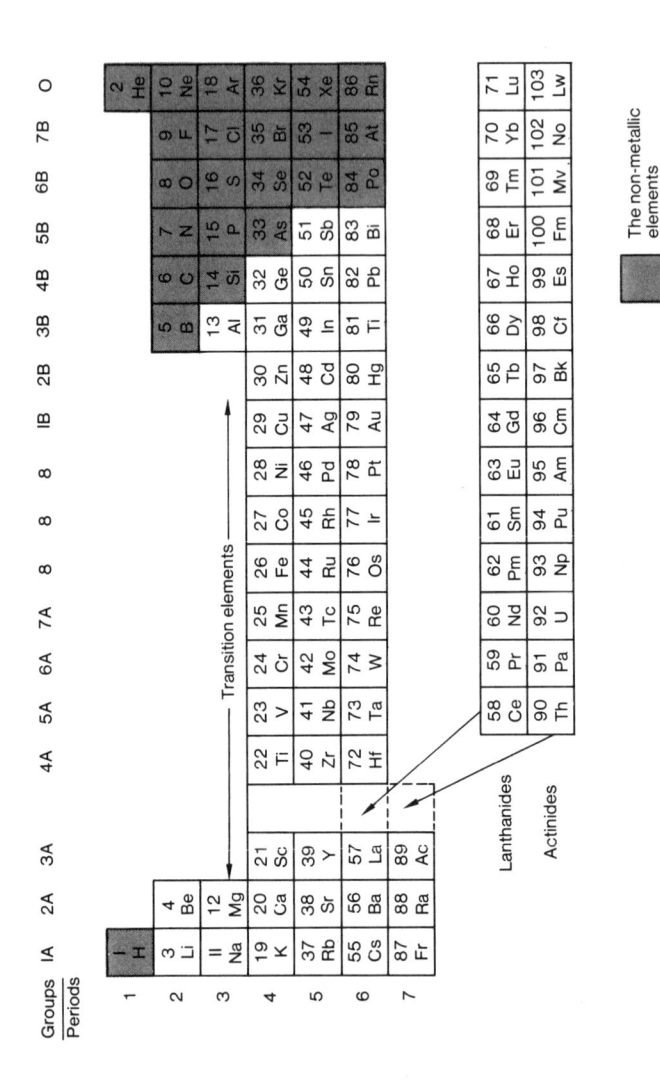

TABLE 1A The Periodic Table

1

THE PERIODIC CLASSIFICATION OF ELEMENTS

1.1 Introduction

Once relative atomic masses (atomic weights) became available in the early nineteenth century, attempts were made to discover if there was any pattern between these figures and the properties of the elements. The most important step in developing a periodic classification was taken in 1869, when the Russian chemist, Mendeléeff, studied the relationship between the relative atomic masses of elements and their properties. He was led to the conclusion that 'the properties of the elements are in periodic dependence on their relative atomic masses'. A modern version of the Periodic Table is shown in Table 1A, the ordering of the elements being based on atomic numbers (the number of protons in the nucleus of the atom) rather than on relative atomic masses.

The table is divided into a number of vertical Groups and 7 horizontal Periods. The Groups 1 to 7 are subdivided into A and B, and Group 8 contains three elements in a given Period. It should not be thought that there is any special resemblance between the sub-groups, except perhaps one of valency; indeed the chemical differences between the sub-groups are often so great that they are best considered separately. Furthermore, there is no special significance in, say, iron, cobalt and nickel being placed in Group 8; these three elements are best treated as members of the first transition series extending from scandium to copper. The division into sub-groups, etc., is merely the method used by Mendeléeff in his original table and there is little need to alter it.

1.2 Some Physical Properties of the Elements

(a) Density

Plot the densities of the solid and liquid elements given in Table 1B (p. 9) against atomic number (go as far as the element bismuth). Study the features of the graph you obtain and relate them as far as possible with the Periodic Table (Table 1A). In particular note what kind of elements appear:

(i) at the minima
(ii) on the ascending portions of the graph
(iii) on the descending portions of the graph.

(b) Melting points

Plot the melting points of the elements given in Table 1B against atomic number (go as far as bromine). Discuss the main features of the graph in relationship to the Periodic Table (Table 1A).

(c) Ionisation energies

The energy required to remove an electron completely from an atom of an element is known as the first ionisation energy; the second ionisation energy is similarly the energy needed to remove completely the second electron from the singly charged ion and so on. The values are quoted in kJ mol^{-1}, and for the potassium atom they are:

$$K \rightarrow K^+ + e^- \quad \text{1st ionisation energy} = 418\cdot4 \text{ kJ mol}^{-1}$$
$$K^+ \rightarrow K^{2+} + e^- \quad \text{2nd ionisation energy} = 3068 \text{ kJ mol}^{-1}$$

Notice the large difference between the two values, due to the fact that it is obviously more difficult to ionise an atom if the atom already bears a positive charge. The simple device below can be used to determine the first ionisation energy of neon (Fig. 1.1).

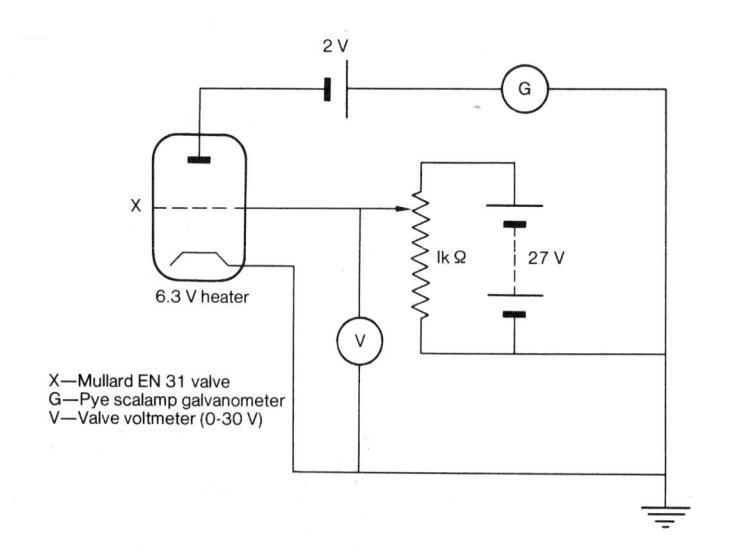

Fig. 1.1
Determination of the first ionisation energy of neon.

TABLE 1B

Element	Density/g cm^{-3}	Melting point/K	First ionisation energy/kJ mol^{-1}	Atomic radius/nm
H		13·9	1313	0·032
He		3·5	2372	
Li	0·54	453·8	520	0·133
Be	1·84	1556	899	0·089
B	2·53	2300	800	0·080
C	2·22*	3700	1091	0·077
N		24·6	1400	0·074
O		54·4	1312	0·074
F		53·6	1679	0·072
Ne		24·6	2080	
Na	0·97	371·0	494	0·157
Mg	1·74	923	738	0·136
Al	2·70	932	578	0·125
Si	2·42	1683	782	0·117
P	1·83*	317·5	1062	0·110
S	2·07*	392·1	1003	0·104
Cl		173	1254	0·099
Ar		83·8	1519	
K	0·87	336·6	418	0·203
Ca	1·54	1123	589	0·174
Sc	3·19	1673	633	0·144
Ti	4·50	1950	661	0·132
V	5·69	2190	649	0·122
Cr	7·10	2176	653	0·117
Mn	7·42	1517	717	0·117
Fe	7·86	1812	762	0·116
Co	8·98	1768	757	0·116
Ni	8·90	1728	736	0·115
Cu	8·94	1356	745	0·117
Zn	7·13	692·7	906	0·125
Ga	5·90	303	578	0·125
Ge	5·46	1210	762	0·122
As	5·73	sublimes 886	946	0·121
Se	4·80	490	941	0·117
Br	3·12	266	1142	0·114
Kr		115·9	1351	
Rb	1·53	312	403	0·216
Sr	2·58	1043	549	0·191
Y	5·51	1773	615	0·162
Zr	6·44	2225	661	0·145
Nb	8·55	2770	664	0·134
Mo	10·20	2890	686	0·129
Tc		2400	703	
Ru	12·53	2700	711	0·124
Rh	12·44	2229	720	0·125
Pd	12·20	1823	803	0·128
Ag	10·50	1234	731	0·134
Cd	8·65	594	868	0·141
In	7·28	429	558	0·150
Sn	7·31	505	708	0·141
Sb	6·70	903	834	0·141
Te	6·25	723	870	0·137
I	4·94	387	1009	0·133
Xe			1170	

* These density values are for graphite, white phosphorus and rhombic sulphur respectively.

A small negative potential of 2 volts is applied to the anode of the valve. Gradually increase the positive potential on the grid until a small deflection is shown by the galvanometer. No current will flow until the potential on the grid is sufficient to produce singly ionised neon Ne^+. The potential in volts is converted into $kJ\ mol^{-1}$ by multiplying by the conversion factor 96·51.

Plot the first ionisation energies of the atoms of the elements given in Table 1B against atomic number. Discuss the main features of the graph in relationship to the Periodic Table (Table 1A).

Can you suggest a reason for the large drop in ionisation energy in going from helium to lithium, neon to sodium, and argon to potassium?

Ignoring the slight irregularities at the positions occupied by beryllium, nitrogen, magnesium and phosphorus, attempt to explain why the ionisation energy rises from lithium to neon and from sodium to argon (refer to the table of atomic radii).

(d) Atomic radii

Atomic radii have been determined using X-ray and electron diffraction techniques. For metals, the inter-nuclear distance in the crystal has been shown to be approximately the same as that for a single metal–metal covalent bond, where compounds containing the latter type of bonds are known. Half the inter-nuclear distance is the atomic radius of the element in question. It is essential to realise that bond lengths in covalent compounds are dependent upon the number of covalent bonds linking two atoms together; for example, there is a decrease in bond length along the series C—C, C=C, C≡C, and compounds containing single covalent bonds must be used to compute atomic radii. For solids, inter-nuclear distances and hence atomic radii do depend to some extent on the way the atoms are arranged in the crystal.

Using the values of atomic radii given in Table 1B, discuss the general trends that occur in:

(i) descending a particular Periodic Group
(ii) traversing a particular Period, e.g. lithium to neon
(iii) traversing the first transition series (scandium to copper).

Is there any obvious correlation between first ionisation energy values and atomic radii?

1.3 Investigation of some Chemical Properties

(a) Action of water on some elements (lithium, magnesium, calcium)

Place some distilled water in a small evaporating dish and then carefully invert in it a test-tube half-full of distilled water. Add a small freshly cut piece of lithium to the water and then quickly place the test-tube over it. Collect the gas given off and show that it is hydrogen by placing the mouth of the test-tube near a Bunsen burner flame (since a hydrogen/air mixture is present in the test-tube a sharp squeak should be heard). Test the resulting solution in the evaporating dish with red litmus paper and then write an equation for the reaction. **The reaction of water with sodium and potassium is violent and should on no account be attempted.**

Repeat the experiment as above using magnesium turnings in place of lithium. What do you observe?

Repeat again with small pieces of calcium.

From the results of these two experiments try to predict what would happen if similar small pieces of barium were added to water. Would you expect the reaction to be more or less violent than with calcium?

(b) Action of steam on magnesium

Set up the apparatus as shown in Fig. 1.2 and make sure that the steam is freely issuing from the hole in the boiling-tube before you heat the magnesium ribbon. Once the magnesium is burning, place a Bunsen burner flame near the hole in the boiling-tube. Hydrogen is evolved which burns.

Find out from a reference book three more metals that react with steam to give hydrogen.

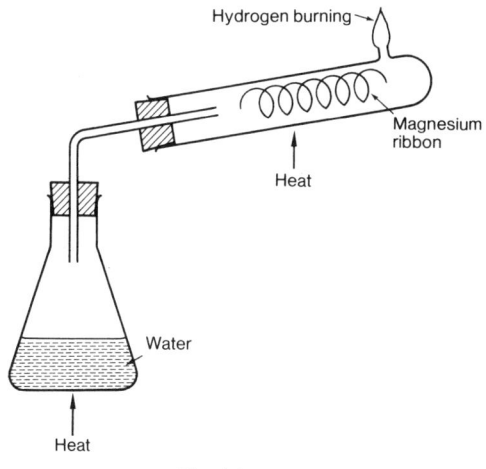

Fig. 1.2
The action of steam on magnesium.

(c) Action of steam on some non-metals (Demonstration)

The non-metals boron, carbon and silicon decompose steam at high temperatures to give hydrogen. The experiment is best carried out in the apparatus shown in Fig. 1.3 (p. 12) which employs an electric furnace.

Allow time for the electric furnace to warm up fully, then heat the water and allow the steam to pass over the hot carbon (granular form). After allowing for the displacement of air, collect some of the gas in a boiling-tube. If the gas burns quietly without exploding it is safe to collect it in gas jars. Apply a lighted splint to a gas jar full of the gas and note the colour of the flame. Add lime water, shake and observe what happens. Do the same test on an unburnt sample of the gas, note and explain the difference this time.

Repeat the experiment with silicon and collect the gas in boiling-tubes. Apply a lighted splint to the gas and observe what happens.

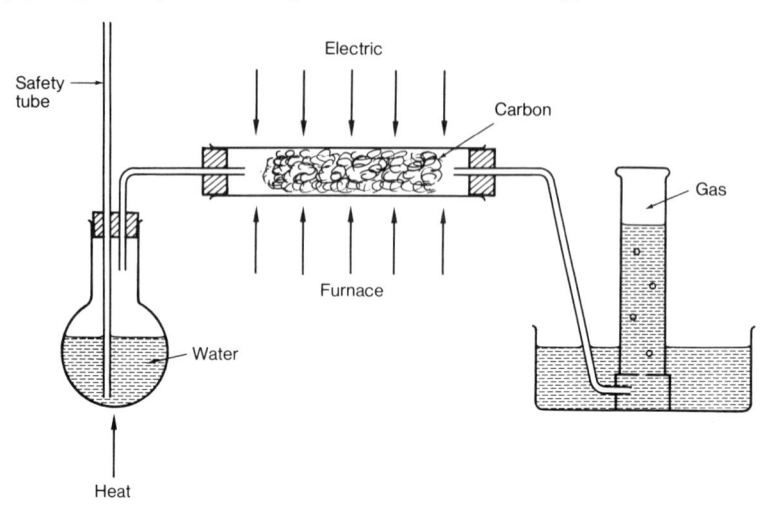

Fig. 1.3
The action of steam on carbon.

(d) Action of hydrochloric acid on some elements

Treat small samples (about 0·5 g) of the following metals separately with approximately 2 M hydrochloric acid: aluminium, bismuth, copper, chromium, iron, lead, magnesium, nickel, tin and zinc and collect any gas evolved as in Fig. 1.4. **Note:** The reaction between magnesium and dilute hydrochloric acid is so rapid that a lighted splint should be applied to the top of the test-tube as soon as the acid has been added to the metal.

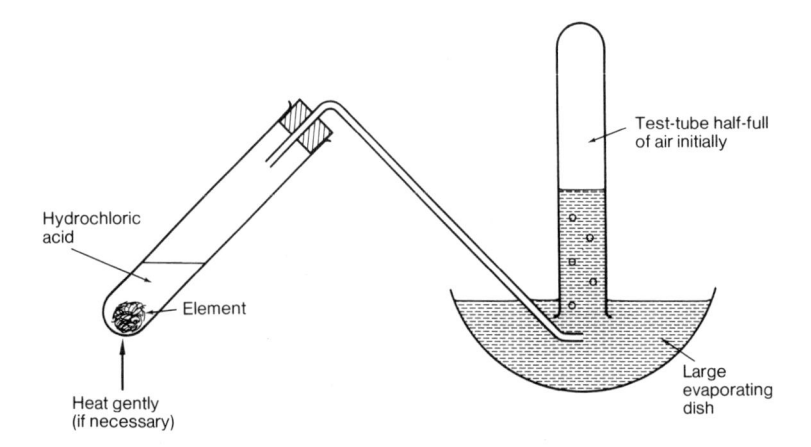

Fig. 1.4
The action of hydrochloric acid on some elements.

If there appears to be little or no reaction in the cold, try warming the test-tube gently; if there is still little or no gas evolved try increasing the concentration of the acid. Make a list of the metals that do and do not evolve hydrogen under these conditions. Also note whether heat has to be applied and whether the acid is dilute or concentrated.

Try the effect of hydrochloric acid on the three non-metals boron, carbon and silicon. Do they evolve any hydrogen under these conditions?

(e) Action of water on some oxides and hydroxides

To about $2\,cm^3$ of distilled water in a test-tube add a small quantity of calcium oxide (0·3 g is sufficient). Now add 2 drops of Universal Indicator solution and note the colour of the indicator.

Repeat the experiment with aluminium oxide (obtained by heating aluminium nitrate until no further change takes place), boron trioxide, iodine(V) oxide, lithium oxide (obtained by heating lithium nitrate until no further change occurs), magnesium oxide, phosphorus(V) oxide (**care: the reaction is violent**), silicon(IV) oxide and zinc oxide. Which of the above oxides give acid and which give alkaline solutions? Can you find any relationship between this behaviour and the position of the various elements in the Periodic Table?

Repeat the experiment with carbon dioxide (obtained by heating copper(II) carbonate) and sulphur dioxide (from a syphon).

Repeat the experiment again, but this time use the following hydroxides: $B(OH)_3$, $Ca(OH)_2$, KOH (one pellet), NaOH (one pellet), $P(OH)_3$. **Caution: KOH and NaOH are caustic to the skin.** Which hydroxides give alkaline and which give acid reactions in water?

(f) Precipitation of metallic hydroxides

You will be provided with a number of aqueous solutions of salts in which the concentration of the particular cation is approximately $0.1\,mol\,dm^{-3}$ (convenient solutions to use are aluminium sulphate, barium chloride, calcium chloride, chromium(III) sulphate, cobalt(II) chloride, copper(II) sulphate, iron(II) sulphate, iron(III) chloride, lead(II) nitrate, manganese(II) sulphate, nickel sulphate, potassium chloride and zinc sulphate). Take about $2\,cm^{-3}$ of each solution in a test-tube and add approximately 2 M sodium hydroxide solution gradually drop by drop. Observe any colour changes, the appearance and colour of any precipitate and whether the precipitate dissolves in an excess of the sodium hydroxide solution. In addition, note any change in colour of the precipitates of cobalt(II) hydroxide and manganese(II) hydroxide on standing. Summarise your results in the form of a table overleaf.

Cation in aqueous solution	Colour of original solution	Colour of the hydroxide precipitate (if any)	Solubility of the precipitate (if any) in excess sodium hydroxide solution
Al^{3+}			
Ba^{2+}			
Ca^{2+}			
Cr^{3+}			
Co^{2+}			
Cu^{2+}			
Fe^{2+}			
Fe^{3+}			
Pb^{2+}			
Mn^{2+}			
Ni^{2+}			
K^{+}			
Zn^{2+}			

Where do the metals that form coloured salts and coloured hydroxide precipitates appear in the Periodic Table? Similarly, where do the metals that form precipitates, which dissolve in an excess of sodium hydroxide solution, appear in the Periodic Table?

All metallic hydroxides dissolve in dilute acid (and this can be confirmed by adding approximately 2 M nitric acid to the precipitates formed above). However, a few metallic hydroxides also dissolve in an excess of sodium hydroxide solution and such hydroxides are said to be amphoteric. This behaviour is easily explained by assuming an equilibrium to be set up in solution; thus if $M(OH)_2$ represents the amphoteric metallic hydroxide:

$$2OH^- + M^{2+} \rightleftharpoons M(OH)_2 + \text{Water} \rightleftharpoons M(OH)_4^{2-} + 2H_3O^+$$
$$\updownarrow 2H_3O^+ \qquad\qquad\qquad\qquad\qquad \updownarrow 2OH^-$$
$$4H_2O \qquad\qquad\qquad\qquad\qquad\qquad 4H_2O$$

The addition of an acid drives the equilibrium to the left as the H_3O^+ ions remove the OH^- ions to form water and the metallic hydroxide $M(OH)_2$ dissolves. Similarly the addition of a strong alkali drives the equilibrium to the right as the OH^- ions combine with the H_3O^+ ions to form water and the metallic hydroxide again dissolves, this time in the form of $M(OH)_4^{2-}$ ions.

(g) Action of water on some hydrides

To about 25 cm³ of distilled water in a small beaker add about 0·2 g of lithium hydride (**caution: the reaction is violent**). Note the evolution of a gas; what do you think it is? Add a few drops of Universal Indicator solution to the resulting solution and note the colour of the solution formed.

Repeat the experiment using calcium hydride.

Mix together about 0·5 g of ammonium chloride and 0·5 g of calcium hydroxide and place the mixture in a test-tube. Warm gently and allow the ammonia gas to come into contact with moist Universal Indicator paper held near the mouth of the test-tube. Note any change in colour of the Universal Indicator paper.

To about 0·5 g of sodium chloride in a test-tube add about 1 cm^3 of concentrated sulphuric acid (**caution: concentrated sulphuric acid is corrosive**). Do the reaction in a fume cupboard and note the evolution of a strongly fuming gas, hydrogen chloride. Allow this gas to come into contact with moist Universal Indicator paper held near the mouth of the test tube and note any change in colour of the Universal Indicator paper.

Repeat the experiment using sodium bromide and concentrated sulphuric acid. This reaction produces the strongly fuming gas hydrogen bromide, together with much free bromine.

From the results obtained in the above experiments with Li^+H^-, $Ca^{2+}(H^-)_2$, NH_3, HCl and HBr, write balanced chemical equations for the reaction of these hydrides with water. Remember that an acidic reaction is due to the presence of H_3O^+ ions and an alkaline reaction is due to OH^- ions.

1.4 Physical Nature of Oxides, Hydrides and Chlorides

The melting points (Kelvin scale) of the oxides, hydrides and chlorides of the first two short periods are plotted in Fig. 1.5 (p. 16). Where a particular element forms more than one oxide, or hydride, or chloride, only one of them is given. The horizontal line drawn at 288 K (average room temperature) clearly shows which compounds are solids (those above the horizontal line) and those which are either liquids or gases (those below the horizontal line). In one or two cases decomposition occurs before the melting-points are reached and the decomposition temperatures are plotted.

Although a full discussion of chemical bonding is beyond the scope of this book, some reference to it is essential.

There are two main types of chemical bonding (excluding the metallic bond), namely electrovalency and covalency. In electrovalency, one or more electrons (rarely more than two) are transferred from the metallic atom to the non-metallic atom. The ions so formed often have the electronic configurations of a noble gas, thus:

$$Na(2.8.1) + Cl(2.8.7) \rightarrow Na^+(2.8) + Cl^-(2.8.8)$$
electronic configuration of neon   electronic configuration of argon

$$Ca(2.8.8.2) + O(2.6) \rightarrow Ca^{2+}(2.8.8) + O^{2-}(2.8)$$
electronic configuration of argon   electronic configuration of neon

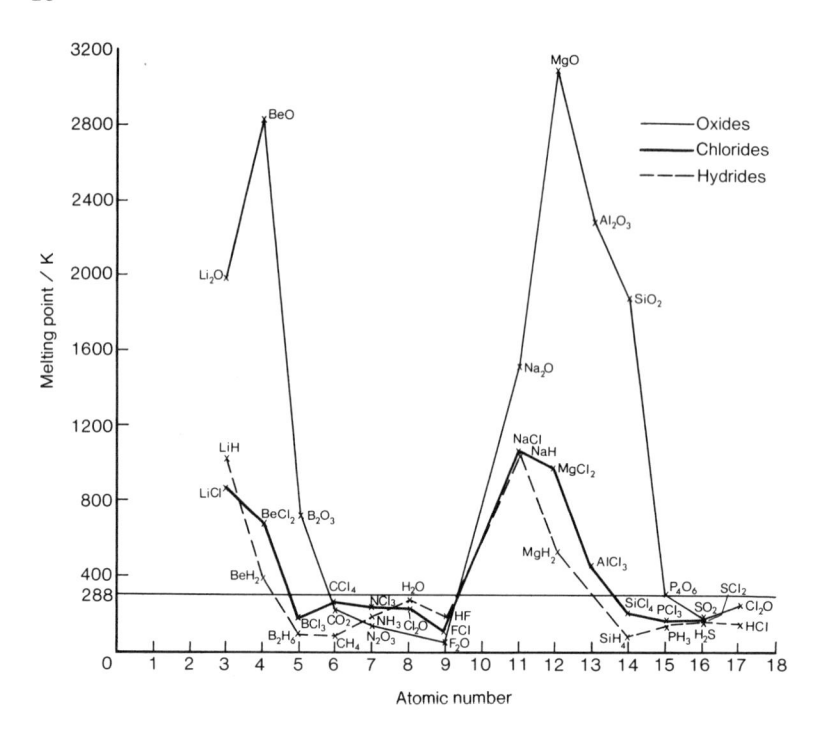

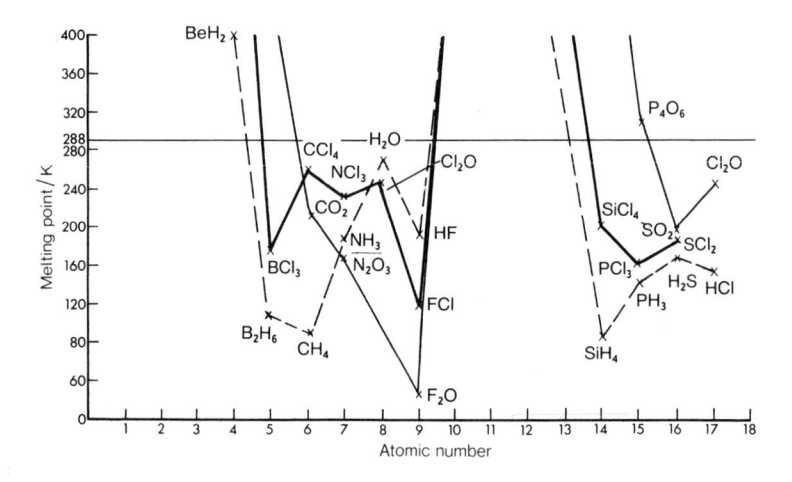

Fig. 1.5
The melting points of the oxides, hydrides and chlorides of elements of the first two short Periods.

Electrovalent compounds exist as hard crystals, of high melting point, with ions of opposite charge arranged in a symmetrical array; thus X-ray examination of sodium chloride shows that each sodium ion is surrounded by six equidistant chloride ions and that each chloride ion is surrounded by six equidistant sodium ions. The actual positions taken up by the ions in a crystal are determined by their charges and relative sizes. In no sense can a single molecule of an electrovalent compound be said to exist, e.g. Na^+Cl^- simply represents the empirical formula of sodium chloride, the whole crystal being one giant molecule or macromolecule.

Electrovalent compounds are formed between the most reactive metals, for example Groups 1A and 2A of the Periodic Table, and the most reactive non-metals, for example Groups 6B and 7B.

A covalent bond is responsible for the bonding together of atoms in simple molecules, the bonded atoms being non-metallic or weakly metallic. Each bonded atom provides one, two or three electrons which are shared, the bonds produced being referred to respectively as single, double and triple covalent bonds. The following are a few examples (the outer electrons only being shown):

$$:\!\ddot{C}l\cdot \; + \; \cdot\ddot{C}l\!: \; \longrightarrow \; :\!\ddot{C}l\!:\!\ddot{C}l\!: \quad \text{or} \quad Cl-Cl$$

$$:\!\ddot{O}\cdot \; + \; \cdot\ddot{O}\!: \; \longrightarrow \; :\!\ddot{O}\!:\!\ddot{O}\!: \quad \text{or} \quad O=O$$

$$\cdot\ddot{N}\cdot \; + \; \cdot\ddot{N}\cdot \; \longrightarrow \; :\!N\!:\!N\!: \quad \text{or} \quad N\equiv N$$

When the two atoms that are bound together by covalent bonds are different, the electrons are not equally shared. For instance, the chlorine atom has a greater affinity for electrons than does the hydrogen atom in hydrogen chloride, and the electron pair constituting the single covalent bond is displaced towards the chlorine atom. The molecule is said to possess an electrical dipole and this is responsible for the substance having a boiling point higher than it would otherwise have had.

$$H\cdot \; + \; {}^{\times}_{\times}\!\overset{\times\times}{\underset{\times\times}{Cl}}{}^{\times}_{\times} \; \longrightarrow \; H{}^{\times}_{\times}\!\overset{\times\times}{\underset{\times\times}{Cl}}{}^{\times}_{\times} \quad \text{or} \quad H-Cl$$

To indicate the presence of an electrical dipole in a molecule, the notation $\overset{\delta+}{A}-\overset{\delta-}{B}$ is often used, e.g. $\overset{\delta+}{H}-\overset{\delta-}{Cl}$.

Three more examples of covalent compounds are given below:

$$2H\cdot + \overset{\times\times}{\underset{\times\times}{\times\ddot{O}\times}} \rightarrow \overset{\times}{\underset{\times}{\times\ddot{O}\times}}\begin{matrix}H\\H\end{matrix} \quad \text{or} \quad O\begin{matrix}H\\H\end{matrix}$$

Water molecule

$$3H\cdot + \overset{\times\times}{\underset{\times}{\times\ddot{N}\times}} \rightarrow \overset{\times}{\underset{\times}{\times N\times}}\begin{matrix}H\\H\\H\end{matrix} \quad \text{or} \quad N\begin{matrix}H\\H\\H\end{matrix}$$

Ammonia molecule

$$\cdot\ddot{C}\cdot + 2\overset{\times\times}{\underset{\times\times}{\times\ddot{O}\times}} \rightarrow \overset{\times}{\underset{\times}{\times\ddot{O}\times}}C\overset{\times}{\underset{\times}{\times\ddot{O}\times}} \quad \text{or} \quad O=C=O$$

Carbon dioxide molecule

The covalent compounds discussed above exist in the form of discrete molecules with little force of attraction between the individual molecules. This is true of many covalent compounds and accounts for many of them being gases, volatile liquids or easily fusible solids. There are, however, some giant molecules or macromolecules, for example silicon dioxide, in which directional covalent bonding extends throughout the whole structure. Compounds of this type are, of course, solids with high melting and boiling points.

As a word of caution, it must be added that electrovalency and covalency are two extremes of chemical bonding, consequently many compounds have bonding of intermediate character. Thus whenever an electron pair is not equally shared between two bonding atoms in a covalent compound, a certain amount of 'ionic character' is introduced, for example, hydrogen chloride is polarised in the sense $\overset{\delta+}{H}\ \overset{\delta-}{Cl}$. Whenever a compound is described as being either electrovalent or covalent, the wording is intended to convey the idea that the bonding is predominantly of one particular type.

Inspection of Fig. 1.5 clearly shows which compounds exist in the form of discrete molecules and which are thus covalent (those below the horizontal line drawn at 288K) and those which are macromolecular (electrovalent or covalent). Typical macromolecular compounds are B_2O_3 (covalent), SiO_2 (covalent), MgO (electrovalent) and NaCl (electrovalent). Aluminium chloride, which appears just above the horizontal line, is covalent and macromolecular but the vapour consists of the dimer Al_2Cl_6. The hydrides of beryllium and magnesium, BeH_2 and MgH_2 respectively, are polymeric but the bonding is certainly neither electrovalent nor covalent. These structures involve hydrogen bridging and are described in detail in some advanced Inorganic texts.

Question: The following pairs of compounds are certainly predominantly covalent but the melting point of the first member (which has a lower relative molecular mass) is higher than that of the second member: NH_3/PH_3, H_2O/H_2S, HF/HCl. Can you give any explanation for this?

1.5 Summary

(a) There is a periodic relationship between the properties of the elements and their atomic numbers, e.g. density, melting point and first ionisation energy.

(b) Metals vary in the ease with which they react with water and acids. Calcium, lithium and sodium, for example, react with water evolving hydrogen. Magnesium reacts with steam, and a large number of metals react with hydrochloric acid with the evolution of hydrogen.

(c) The non-metals carbon and silicon react with steam when heated to a high temperature with the formation of hydrogen. No non-metal reacts with hydrochloric acid.

(d) The oxides of reactive metals, e.g. calcium oxide and lithium oxide react with water to give alkaline solutions. Magnesium oxide reacts slightly to give an alkaline reaction. This type of behaviour is typical of metals in Groups 1A and 2A of the Periodic Table. The hydroxides of these metals also give an alkaline reaction in water.

(e) The oxides of many non-metals, e.g. boron oxide, carbon dioxide, phosphorus(V) oxide, sulphur dioxide and iodine(V) oxide, react with water to give acidic solutions. These oxides are those of non-metals which occur in Groups 3B, 4B, 5B, 6B and 7B of the Periodic Table. The 'hydroxides' of non-metals, e.g. $B(OH)_3$ and $P(OH)_3$, also give acidic solutions with water; they are more generally called acids.

(f) Many metals form insoluble hydroxides. Transition metals form coloured hydroxides, while the hydroxides of aluminium, lead(II) and zinc dissolve in an excess of sodium hydroxide solution and are called amphoteric hydroxides.

(g) The hydrides of Groups 1A and 2A metals are ionic and react with water to give alkaline solutions, at the same time liberating hydrogen.

(h) The hydrides of non-metals are covalent, e.g. NH_3, HBr and HCl. Ammonia gives an alkaline solution with water, whereas hydrogen bromide and hydrogen chloride give strongly acidic solutions.

2

GROUP 1A THE ALKALI METALS

2.1 Some Physical Data of Group 1A Elements

	Atomic number	Electronic configuration	First ionisation energy/ kJ mol^{-1}	Density/ g cm^{-3}	M.p./ K	B.p./ K	Atomic radius/ nm	Ionic radius/ nm
Li	3	2.1 $1s^22s^1$	520	0.54	454	1613	0·133	0·060
Na	11	2.8.1 ...$2s^22p^63s^1$	494	0.97	371	1158	0·157	0·095
K	19	2.8.8.1 ...$3s^23p^64s^1$	418	0.87	337	1048	0·203	0·133
Rb	37	2.8.18.8.1 ...$4s^24p^65s^1$	403	1·53	312	963	0·216	0·148
Cs	55	2.8.18.18.8.1 ...$5s^25p^66s^1$	374	1·90	302	943	0·235	0·169

2.2 Some General Remarks about Group 1A

The elements lithium, sodium, potassium, rubidium, caesium and francium are called the alkali metals. Not much is known about the last-named since it is radioactive and all its isotopes are exceedingly short-lived; it is formed during the decay of actinium.

The metals are extremely reactive and electropositive (prone to lose electrons) and exist in combination with other elements or radicals as positive ions, e.g. Na^+Cl^-, $(K^+)_2SO_4^{2-}$; they are therefore never found in the free state in nature. They all adopt the body-centred cubic structure in which each atom is surrounded by eight nearest neighbours with six more atoms only slightly further distant (Fig. 2.1).

All alkali metal atoms have one electron in the outer shell, preceeded by a closed shell containing eight electrons—except lithium which has a closed shell of two. In chemical combination this single electron is transferred to a non-metallic atom, giving a unipositive metal ion with the stable electronic configuration of a noble gas, e.g. $Na^+(2·8)$ is isoelectronic with $Ne(2·8)$. Compounds of the alkali metals are therefore generally predominantly ionic and exist as high melting-point solids in which as many ions of opposite charge surround each other as possible.

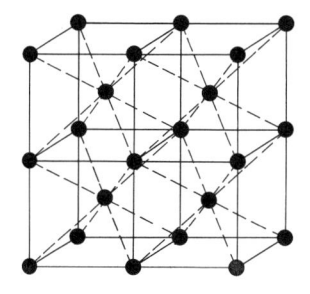

Fig. 2.1
The body-centred cubic structure of the alkali metals.

Although the alkali metals form predominantly ionic compounds, they can form covalent molecules such as Li_2, Na_2, K_2, etc. which are found to the extent of about 1 per cent in the vapours of these metals. Covalent compounds such as lithium phenyl, C_6H_5Li, exist but they tend to be highly reactive.

2.3 Extraction of the Group 1A Metals

These metals are extracted by electrolysis of their fused chlorides, other halides being added to lower their melting points and thus economise on electrical power.

(a) Electrolysis of fused lithium chloride (Demonstration)

Two-thirds fill a nickel crucible with anhydrous lithium chloride and place it in a pipeclay triangle. Fuse the salt with a hot Bunsen flame and then insert a steel cathode and a graphite anode completing the circuit with an ammeter, a variable resistance and a source of about 12 V DC. Adjust the variable resistance until a current of about 4 A is flowing, turn down the Bunsen flame so that the lithium chloride is kept just molten and continue to electrolyse for about 10 minutes (Fig. 2.2, p. 22). Switch off the current, remove the Bunsen burner and the electrodes and allow the cathode to cool. Carefully dip the cathode into a beaker containing distilled water and note the evolution of gas. What is it? Add Universal Indicator solution to the beaker and note its colour. What is produced in solution? Write an equation for the reaction of lithium with water.

(b) Electrolysis of fused sodium chloride (Demonstration)

Carry out the same experiment but substitute sodium chloride for the lithium chloride. It is now necessary to mix some anhydrous calcium chloride with the sodium chloride (about 1% by mass) in order to depress the melting-point of the sodium chloride, c.f. melting-points of lithium chloride and

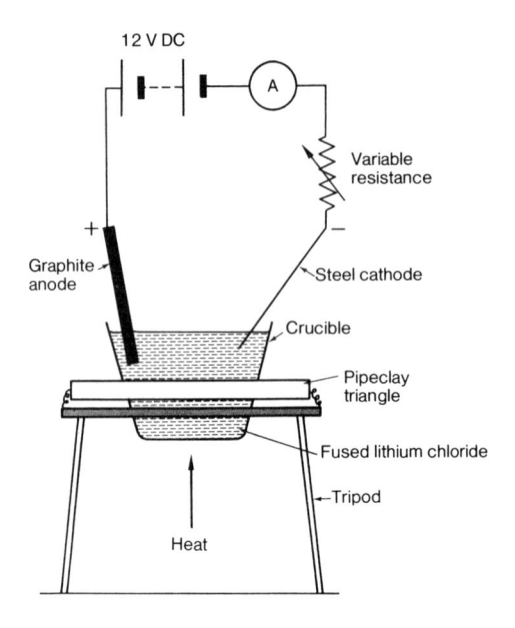

Fig. 2.2
Electrolysis of fused lithium chloride.

sodium chloride (Fig. 1.5, p. 16); it is best to substitute a blowlamp for the Bunsen burner. Sodium is liberated at the cathode and generally catches fire in a series of yellow flashes.

2.4 Some Reactions of the Group 1A Metals

(a) Burning lithium and sodium in oxygen

Fill two boiling-tubes with oxygen from a cylinder and cork the two tubes until they are used. Cut a clean piece of lithium (a cube of side about 2 mm) and place it in a clean combustion spoon. Heat the lithium in a Bunsen flame until it is burning and then transfer the combustion spoon to the boiling-tube full of oxygen. Record the colour of the burning lithium and note the intensity of the reaction. The product is lithium oxide, $(Li^+)_2O^{2-}$, which is white when pure.

Repeat the experiment using a similar sized piece of clean sodium. Again note the colour of the burning metal. This time a mixture of sodium monoxide, $(Na^+)_2O^{2-}$, and sodium peroxide, $(Na^+)_2O_2^{2-}$, is formed.

The other Group 1A metals give the superoxide when reacted with oxygen, e.g. potassium forms potassium superoxide, $K^+O_2^-$.

(b) Burning lithium and sodium in chlorine

Fill two boiling-tubes with chlorine **(in a fume cupboard)** and cork the two tubes until they are used. Cut a clean piece of lithium (a cube of side about 2 mm) and place it in a clean combustion spoon. Heat the lithium in a Bunsen flame until it is just on the point of burning and then transfer the combustion spoon to the boiling-tube full of chlorine. Note the vigour of the combination of lithium with chlorine. The product is lithium chloride, Li^+Cl^- which is white when pure.

Repeat the experiment with a similar sized piece of sodium. Again note the vigour of the reaction. Sodium chloride, Na^+Cl^-, is formed.

The other Group 1A metals react with chlorine in a similar manner.

(c) Action of water on lithium

Place some distilled water in a small evaporating dish and then carefully invert in it a test-tube half-full of distilled water. Add a small freshly cut piece of lithium to the water and then quickly place the test-tube over it. Collect the gas given off and show that it is hydrogen by placing the mouth of the test-tube near a Bunsen burner flame (since a hydrogen/air mixture is present in the test tube a sharp squeak should be heard). Test the resulting solution with Universal Indicator solution or litmus solution and then write an equation for the reaction of lithium with water.

Sodium and potassium react in a similar manner but the reactions become increasingly violent and **should on no account be attempted.** In the case of potassium, sufficient heat is generated to ignite the hydrogen and the metal, which burn with a lilac flame (flame test for potassium compounds see p. 26).

2.5 Oxides and Hydroxides of the Group 1A Metals

Lithium only forms the monoxide $(Li^+)_2O^{2-}$ when heated in oxygen, sodium forms the monoxide, and the peroxide $(Na^+)_2O_2^{2-}$ if an excess of oxygen is used. The other Group 1A metals give the superoxide when reacted with oxygen, e.g. $K^+O_2^-$. The lithium ion, by virtue of its small size, is not able to surround itself by sufficient peroxide ions to give a stable crystal lattice and consequently only the monoxide exists. The ions of potassium, rubidium and caesium get progressively larger in this order and are able to form stable structures with the superoxide ion—the largest of the three oxide ions. These ions are related as follows:

$$O^{2-} \xrightarrow{\frac{1}{2}O_2} O_2^{2-} \xrightarrow{O_2} 2O_2^-$$

(a) Action of water on lithium monoxide and sodium peroxide

Heat a small amount of lithium nitrate (about 0.5 g) strongly in an ignition tube until no further change occurs; nitrogen dioxide and oxygen are evolved (test with a glowing splint). The residue is lithium monoxide:

$$4Li^+NO_3^- \rightarrow 2(Li^+)_2O^{2-} + 4NO_2 + O_2$$

Add about 2 cm³ of distilled water to the residue after cooling and then add 2 drops of Universal Indicator solution. Write an equation for the reaction of water with lithium monoxide.

Sodium peroxide, $(Na^+)_2O_2^{2-}$, and potassium superoxide, $K^+O_2^-$, are also unstable in the presence of water and react as follows:

$$(Na^+)_2O_2^{2-} + 2H_2O \rightarrow 2Na^+OH^- + H_2O_2$$

or

$$O_2^{2-} + 2H_2O \rightarrow 2OH^- + H_2O_2$$

$$2K^+O_2^- + 2H_2O \rightarrow 2K^+OH^- + H_2O_2 + O_2$$

or

$$2O_2^- + 2H_2O \rightarrow 2OH^- + H_2O_2 + O_2$$

Cautiously add about 0·1 g of sodium peroxide to about 3 cm³ of distilled water in a test-tube. Shake the mixture and show that the resulting solution is alkaline by testing with Universal Indicator solution.

(b) Action of water on lithium hydroxide, sodium hydroxide and potassium hydroxide

Caution: these hydroxides are caustic and should on no account be touched with the fingers.

Into three separate test-tubes place respectively about 0·5 g of lithium hydroxide, one pellet of sodium hydroxide and one pellet of potassium hydroxide. Add about 2 cm³ of distilled water to each test-tube, stir and insert a thermometer. Comment on the results.

Into three separate watch glasses place about 0·5 g of lithium hydroxide, one pellet of sodium hydroxide and one pellet of potassium hydroxide. Examine over a period of about one hour and then take a final look after about 24 hours. Comment on your observations.

(c) Formation of sodium hydroxide by electrolysis of an aqueous solution of sodium chloride (Demonstration)

Commercially, sodium hydroxide is obtained by the electrolysis of an aqueous solution of sodium chloride using a mercury cathode (Castner–Kellner process). For details, a text book should be consulted. Potassium hydroxide is manufactured by a similar process from potassium chloride solution.

In view of the toxic properties of mercury vapour it is best to restrict the use of this element in the laboratory and the following demonstration uses a steel cathode. The sodium hydroxide solution obtained is, however, contaminated with sodium chloride.

Set up the apparatus shown in Fig. 2.3 using a saturated solution of sodium chloride as the electrolyte. After electrolysing for several minutes add a few drops of Universal Indicator solution or some phenolphthalein

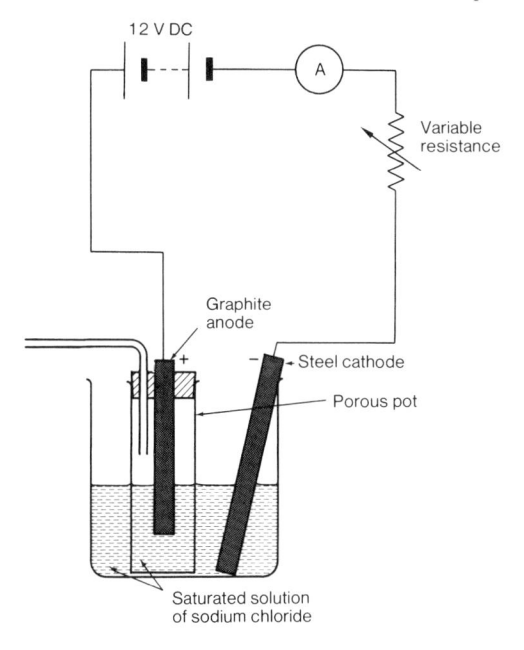

Fig. 2.3
Electrolysis of aqueous sodium chloride solution.

solution to the solution into which the steel cathode is dipping (cathode compartment) and note its colour. Note that a gas is evolved at the cathode. What do you think it is? Hold a piece of filter paper previously dipped into potassium iodide solution near the mouth of the tube leading from the porous pot (anode compartment). What happens? Can you explain this observation? Having recorded all your observations, attempt to explain all the features of this electrolysis; consult a text book of inorganic or physical chemistry to confirm your answers.

2.6 The Hydrides of the Group 1A Metals

All the Group 1A metals react with hydrogen on heating to give a hydride of the general formula M^+H^-. Proof that these hydrides contain the negative hydrogen ion, H^-, has been obtained by electrolysis, e.g. electrolysis of fused lithium hydride results in hydrogen being liberated at the anode.

Action of water on lithium hydride

To about 25 cm^3 of distilled water in a small beaker add about 0·2 g of lithium hydride (**caution: the reaction is violent**). Note the evolution of a gas; what do you think it is? Add a few drops of Universal Indicator solution to the resulting solution, note its colour and explain the result. Write an equation for the reaction of lithium hydride with water.

The hydrides of the other Group 1A metals react in a similar manner.

2.7 Flame Tests for the Group 1A Metal Compounds

All the Group 1A metals give characteristic flame colorations which can be used to identify them in compounds. These flame colours are caused by an electron or electrons previously excited to a higher energy level by the Bunsen flame falling back into a lower energy level. The energy released in these particular instances results in radiation being emitted in the visible part of the electromagnetic spectrum.

(a) Flame tests for lithium, sodium and potassium compounds

Clean a nichrome wire by dipping it into some concentrated hydrochloric acid, contained in a watch glass, and then holding it in the Bunsen flame. Repeat the procedure until the wire gives no flame coloration. Now dip the wire into some fresh acid and then into some powdered lithium chloride (or nitrate). Hold the moistened solid, sticking onto the nichrome wire, in the Bunsen flame. Record the flame colour.

Repeat the experiment with sodium chloride and potassium chloride (preferably analytical reagent grade, since sodium compound impurities tend to mask the colour given by potassium compounds). Record the flame colorations.

(b) Flame tests for rubidium and caesium compounds (Demonstration)

In view of the expense of these compounds, these tests are best done as a demonstration. Repeat the procedure above using the chlorides and record the flame colorations.

If a direct vision spectroscope is available view the flame colorations with this instrument, since more than one coloured spectral line may be present. For example, the characteristic colour given by lithium compounds is due to two spectral lines.

2.8 Thermal Stability of Group 1A Metal Salts

(a) Action of heat on the carbonates of lithium, sodium and potassium

Place about 0·5 g of lithium carbonate in an ignition tube and heat, gently at first and then more strongly. Pass any carbon dioxide evolved into a test-tube containing about 2 cm³ of lime water. With the thumb over the test-tube, shake the lime water and note whether it turns milky.

Repeat the experiment with sodium carbonate and potassium carbonate (both anhydrous and reagent grade). Can you say which of these three carbonates is the most thermally unstable?

(b) Action of heat on the nitrates of lithium, sodium and potassium

Place about 0·5 g of lithium nitrate in an ignition tube and heat strongly. Once all the water of crystallisation has been driven off test for oxygen with a glowing splint. What other gas is evolved? Show that it is an acidic gas by testing with moist Universal Indicator paper.

Repeat the experiment with sodium nitrate and potassium nitrate. Is any oxygen evolved on strong heating? Is any other gas evolved? Which of the three nitrates is the most thermally unstable?

(c) Action of heat on the hydrogen carbonates of sodium and potassium

Place about 0·5 g of anhydrous sodium hydrogen carbonate in an ignition tube and heat gently and test for the evolution of carbon dioxide in the usual way. Note any other changes that occur. Repeat the experiment with anhydrous potassium hydrogen carbonate.

To the cool residues add about 2 cm³ of approximately 2M hydrochloric acid and test for the evolution of carbon dioxide.

Taking into account your observations on the action of heat on sodium carbonate and potassium carbonate, try to write an equation for the action of mild heat on these two hydrogen carbonates, $M^+HCO_3^-$.

Lithium hydrogen carbonate is not listed in any chemical catalogue. In view of what you have discovered in Experiments 2.8 (a), (b) and (c) can you offer any explanation?

2.9 Appearance and Solubility of Salts of the Group 1A Metals

(a) General appearance

Note the colours of a large range of salts of sodium and potassium and of as many lithium salts as possible, e.g. in addition to the ones you have already handled include the bromide, chromate, iodide, nitrite, permanganate, phosphates and sulphate of sodium and potassium. Which salts are coloured? What ions do you think are responsible for these colours?

(b) Solubility of some Group 1A metal salts

The vast majority of salts of sodium and potassium are appreciably soluble in water. Lithium, however, has several salts that are only sparingly soluble in water, for example, the fluoride and phosphate.

Precipitation of lithium fluoride

To about 3 cm³ of an approximately 0·2M solution of lithium chloride add 3 cm³ of an approximately 0·2M solution of sodium fluoride and then about 2 drops of 2M aqueous ammonia to ensure an alkaline solution. Note the formation of a white precipitate of the sparingly soluble lithium fluoride, Li^+F^-.

2.10 Lithium Chloride has Some Covalent Character

Many lithium compounds display appreciable covalent character and this manifests itself by solubility in what are called 'organic solvents', for example, methanol. Can you explain why lithium should display appreciable covalent character (look at the physical data given in Section 2.1)?

Action of water and methanol on the chlorides of lithium, sodium and potassium

Place about 0·5 g of lithium chloride in a test-tube and then add about 3 cm³ of distilled water. Insert a thermometer and note any temperature change. Repeat the experiment with sodium chloride and potassium chloride. Comment on your observations.

Repeat the above but this time replace the 3 cm³ of water by the same volume of methanol. Record and comment on your observations.

2.11 Summary

(a) The Group 1A metals are very reactive (the reactivity increases from top to bottom in the Group) and the only economic method of extraction involves electrolysis of their fused chlorides.

(b) The metals react vigorously with non-metals such as chlorine and oxygen to give white ionic compounds, and with water to give alkaline solutions and hydrogen.

(c) The complexity of the oxides formed increases with increasing size of the metallic cation, e.g. lithium only forms the monoxide, $(Li^+)_2O^{2-}$, while sodium forms both a monoxide and a peroxide, $(Na^+)_2O_2^{2-}$, and potassium forms a superoxide, $K^+O_2^-$. These oxides react with water to give alkaline solutions; oxygen is evolved as well from a peroxide and a superoxide.

(d) The hydrides of the Group 1A metals are ionic solids and contain the H^- anion. They react vigorously with water to give alkaline solutions and hydrogen.

(e) Compounds of the Group 1A metals give coloured flame tests—lithium (red), sodium (yellow), potassium (lilac), rubidium (red) and caesium (blue).

(f) The thermal stabilities of salts containing a common anion increase from top to bottom in the Periodic Group, e.g. lithium carbonate is decomposed fairly readily on heating but potassium carbonate is fairly resistant towards heat.
(g) The majority of the salts of the Group 1A metals are white unless the anion happens to be coloured; for example, the chromates are yellow and the permanganates purple. The majority of their salts are soluble in water, although lithium has some fairly insoluble ones, for example, lithium fluoride and lithium phosphate.
(h) Many lithium compounds display appreciable covalent character, unlike most compounds of the other Group 1A metals; for example, lithium chloride is soluble in methanol (an indication of covalent character) whereas sodium chloride is not.

3

GROUP 2A THE ALKALINE EARTH METALS

3.1 Some Physical Data of Group 2A Elements

	Atomic number	Electronic configuration	Ionisation energy/ kJ mol^{-1} First	Second	Density/ g cm^{-3}	M.p./ K	B.p./ K	Atomic Radius/ nm	Ionic radius/ nm
Be	4	2.2 $1s^22s^2$	899	1757	1.86	1556	3240	0·089	0·031
Mg	12	2.8.2 ...$2s^22p^63s^2$	738	1451	1·75	923	1370	0·136	0·065
Ca	20	2.8.8.2 ...$3s^23p^64s^2$	589	1145	1·55	1123	1760	0·174	0·097
Sr	38	2.8.18.8.2 ...$4s^24p^65s^2$	549	1064	2·60	1043	1650	0·191	0·113
Ba	56	2.8.18.18.8.2 ...$5s^25p^66s^2$	502	965	3·60	980	1910	0·198	0·135
Ra	88	2.8.18.32.18.8.2 ...$6s^26p^67s^2$			5.0	970	1410		

3.2 Some General Remarks about Group 2A

Beryllium differs considerably in its chemistry from the other members of this Group, and its compounds, when anhydrous, show a considerable degree of covalent character—a consequence of the small size of the beryllium atom. Magnesium, too, forms compounds which show appreciable covalent character. The other four members of this Group, referred to as 'the alkaline earths', have a very similar chemistry and their compounds are essentially ionic. Radium is radioactive and is formed during the decay of $^{238}_{92}U$ compounds.

All members of this Group are highly reactive and are thus never found in the free state in nature. Beryllium and magnesium have the hexagonal close-packed structure, and strontium the cubic close-packed structure; these two basic structures are two ways in which spheres of equal radius can be close packed and the co-ordination number of each sphere is twelve (see a text book of inorganic chemistry for a fuller discussion). Calcium can adopt either of these two basic structures but barium has the body-centred structure of the alkali metals (p. 21) in which the co-ordination number is eight.

The Group 2A metal atoms have two electrons in the outer shell preceded by a closed shell containing eight electrons—except beryllium which has a closed shell of two. In chemical combination these two outer electrons are transferred, giving a dipositive metal ion with the stable electronic configuration of a noble gas, e.g. Mg^{2+}, which is isoelectronic with $Na^+(2.8)$ and $Ne(2.8)$. Compounds formed by the last four members of this Group—calcium, strontium, barium and radium—are therefore predominantly ionic and exist as high melting-point solids in which as many ions as possible of opposite charge surround each other. The compounds of magnesium are generally predominantly ionic, but, as mentioned above, they do show some covalent character, i.e. an incomplete transfer of the two outer electrons during chemical combination. This trend is even more pronounced with beryllium and its chloride, for example, $BeCl_2$ is considered to be essentially covalent, since it shows very little electrical conductivity in the fused state.

The Group 2A metals have higher melting and boiling-points and higher densities than corresponding metals in Group 1A; this is because the former have two and the latter only one valency electron per atom for bonding the atoms into a metallic structure.

This Group of metals forms more covalent compounds than those of Group 1A, particularly beryllium and magnesium. The highly reactive Grignard reagents which are covalent, e.g. C_2H_5MgBr, are important reagents in organic chemistry.

3.3 Extraction of the Group 2A Metals

Like the metals of Group 1A, these metals are extracted by electrolysis of their fused halides, other halides being added to lower the melting-point and, in the case of beryllium, to make the melt a better electrical conductor.

3.4 Some Reactions of Magnesium and Calcium

(a) Burning magnesium and calcium in oxygen

Fill a boiling-tube with oxygen and then cork it until required. Wrap a small piece of magnesium ribbon (about 4 cm in length) round a combustion spoon and then ignite the ribbon in the Bunsen flame. Immediately place the burning magnesium in the boiling-tube full of oxygen, **taking care not to look directly at the burning magnesium.** Note the intensity of the reaction and the colour of the product, magnesium oxide, $Mg^{2+}O^{2-}$.

It is difficult to initiate the reaction between calcium pellets and oxygen using the procedure as above. Therefore, place one or two pellets of calcium on a gauze and direct a Bunsen burner flame directly onto the calcium (**caution: this reaction can be sudden and very vigorous**). Note the intensity and colour of the burning calcium, and the colour of the product, calcium oxide, $Ca^{2+}O^{2-}$.

The other Group 2A metals react in a similar manner, but strontium and barium also form the peroxide, $M^{2+}O_2^{2-}$, on prolonged heating.

(b) Burning magnesium in chlorine

Fill a boiling-tube with chlorine (**in a fume cupboard**) and cork it until required. Wrap a small piece of magnesium ribbon (about 4 cm in length) round a combustion spoon and then ignite the ribbon. Immediately place the combustion tube in the boiling-tube containing chlorine, and note the vigour of the reaction. The product is magnesium chloride, $Mg^{2+}(Cl^-)_2$, which is very deliquescent and white when pure.

Calcium reacts in a similar manner forming calcium chloride, $Ca^{2+}(Cl^-)_2$, but the reaction between calcium pellets and chlorine is difficult to initiate using the simple apparatus as above. Like magnesium chloride, calcium chloride is very deliquescent and white when pure.

The other alkaline earth metals react with chlorine in a similar manner.

(c) Reaction of magnesium and calcium with nitrogen

Magnesium and calcium react with nitrogen on heating to give the nitride, $(M^{2+})_3(N^{3-})_2$, (see section 3.7 p. 34).

(d) Action of water on magnesium and calcium

Place about 0·2 g of magnesium turnings in a test-tube and then add about 3 cm³ of distilled water. Leave for a while and then examine to see if you can detect any gas being evolved. What is your conclusion? (For the action of steam on magnesium see Section 1.3 p. 11).

Place some distilled water in a small evaporating dish and then carefully invert in it a test-tube half-full of distilled water. Add a few pellets of calcium to the water and then quickly place the test-tube over them. Collect the gas given off, and show that it is hydrogen by placing the mouth of the test-tube near a Bunsen burner flame (since a hydrogen/air mixture is present in the test tube a sharp squeak should be heard). Test the resulting solution with Universal Indicator solution and then write an equation for the reaction of calcium with water.

Why do you think the liquid goes milky after calcium has been reacting with the water for some time?

(e) Action of dilute hydrochloric acid on magnesium and calcium

Place about 0·2 g of magnesium turnings in a test-tube and add about 3 cm³ of approximately 2M hydrochloric acid. Immediately hold the mouth of the test-tube near a Bunsen burner flame. What happens and why?

Repeat the experiment with 2 or 3 pellets of calcium (**no more, since the reaction is violent**) and again place the mouth of the test-tube near a Bunsen burner flame. What happens and why?

Predict how strontium and barium would react with dilute hydrochloric acid.

3.5 Oxides and Hydroxides of the Group 2A Metals

Beryllium, magnesium and calcium form the normal oxide, $M^{2+}O^{2-}$, on heating in oxygen, but with strontium and barium some peroxide, $M^{2+}O_2^{2-}$, is formed as well. Can you explain why?

(a) Action of water on the Group 2A oxides

Place in separate test-tubes about 0·2 g of the following oxides: magnesium oxide, calcium oxide, strontium oxide and barium oxide, then add about 2 cm³ of distilled water to each and a few drops of Universal Indicator solution. Note the results. Do all the oxides react with water to give an alkaline solution?

(b) Action of water on the Group 2A hydroxides

Repeat Experiment 3.5(a) but use the hydroxide in place of the oxide. Are the hydroxides sufficiently soluble to give an alkaline reaction? Would you say the hydroxides were very soluble, sparingly soluble or nearly insoluble in water?

(c) Action of heat on barium peroxide

Place about 0·5 g of barium peroxide in an ignition tube and heat gently and then more strongly if necessary. Test for oxygen with a glowing splint. Write an equation for the decomposition of barium peroxide.

Would you expect strontium peroxide to decompose more readily or less readily on heating?

(d) Relative solubilities of the hydroxides in water

For this experiment you will require aqueous solutions which are approximately 0·5M with respect to Mg^{2+}, Ca^{2+}, Sr^{2+} and Ba^{2+}, prepared from either the metallic chloride or the nitrate. In addition you will need approximately 0·05M barium hydroxide solution (in a burette fitted with a soda-lime trap to prevent attack by atmospheric carbon dioxide) and carbonate-free aqueous ammonia (obtained by diluting 0·880 ammonia four times with distilled water when required).

To about 3 cm³ of each of the aqueous solutions of the Group 2A cations in separate test-tubes add an equal volume of barium hydroxide solution from the special burette. Place the test-tubes in a rack and observe which hydroxides are precipitated (the precipitate may be heavy or only faint, so observe the tubes carefully). Record which hydroxides are sufficiently soluble under these conditions not to precipitate, those which form a faint precipitate and those which form a heavy precipitate.

Repeat the experiments but this time substitute the aqueous ammonia solution for the barium hydroxide solution. Comment on your results and try to explain any differences.

From both sets of results is it possible to arrange the hydroxides in increasing order of solubility?

Would you expect radium hydroxide to be more soluble or less soluble in water than barium hydroxide?

3.6 The Hydrides of the Group 2A Metals

Beryllium and magnesium hydrides are polymeric with structures intermediate between ionic and covalent. The other Group 2A hydrides are ionic, e.g. $M^{2+}(H^-)_2$, and the fact that they contain the negative hydride ion is proved by electrolysis in fused halide solution when hydrogen is evolved at the anode.

Action of water on calcium hydride

To about 25 cm³ of distilled water in a small beaker add about 0·2 g of calcium hydride (**caution: the reaction is violent**). Note the evolution of a gas; what do you think it is? Add a few drops of Universal Indicator solution to the resulting mixture and explain the result. Write an equation for the reaction of calcium hydride with water.

Strontium and barium hydrides behave in a similar manner.

3.7 The Nitrides of the Group 2A Metals

All the Group 2A metals form a nitride of the general formula $(M^{2+})_3(N^{3-})_2$ on sufficient heating in nitrogen. The nitrides of magnesium and calcium are prepared below by heating the metals in air (oxides formed as well).

All the Group 2A nitrides react in a similar way with water.

(a) Formation of magnesium and calcium nitrides

Heat about 0·2 g of magnesium turnings in a crucible and continue to heat until all the magnesium has reacted. The residue is a mixture of magnesium oxide and magnesium nitride.

Repeat with 0·2 g of calcium pellets; the residue is a mixture of calcium oxide and calcium nitride.

(b) Action of water on magnesium and calcium nitrides

Place the cool residue from (a) above in a test-tube and add about 3 cm³ of distilled water. Heat gently, holding a moist piece of Universal Indicator paper (or moist red litmus paper) near, but not touching, the mouth of the test-tube. What gas is evolved? Attempt to write an equation for the action of water on magnesium nitride.

Repeat the experiment with calcium nitride. Is the action of water on this compound similar to its action on magnesium nitride?

3.8 Flame Tests for the Group 2A Metal Compounds

Carry out the flame tests on the chlorides of magnesium, calcium, strontium and barium as explained in Section 2.7 p. 26 and record the flame colorations.

If a direct vision spectroscope is available, view the flame colorations with this instrument and sketch the emission lines. Check with the positions given in a reference book.

3.9 Thermal Stability of Group 2A Metal Salts

(a) Action of heat on the carbonates of magnesium, calcium, strontium and barium

Place 0·5 g of magnesium carbonate in an ignition tube and heat, gently at first and then more strongly. Pass any carbon dioxide evolved into a test-tube containing about 2 cm³ of lime water. With the thumb over the test-tube, shake the lime water and note whether it turns milky.

Repeat the experiment with calcium, strontium and barium carbonates. What can you deduce about the thermal stability of these four carbonates?

(b) Action of heat on the nitrates of magnesium, calcium, strontium and barium

Place about 0·5 g of magnesium nitrate in an ignition tube and heat strongly. Once all the water of crystallisation has been driven off, test for oxygen with a glowing splint. What other gas is evolved?

Repeat with calcium, strontium and barium nitrates. Do they behave in a similar manner? How does the action of heat on these nitrates differ from the action of heat on sodium and potassium nitrates?

(c) Action of heat on the hydrated chlorides of magnesium, calcium, strontium and barium

Place 0·5 g of hydrated magnesium chloride in an ignition tube and heat strongly. Hold a piece of moist Universal Indicator paper (or moist blue litmus paper) near the mouth of the tube and observe what happens. What is the gas evolved?

Continue heating until no further change occurs and then allow the tube to cool. Add about $3\,\text{cm}^3$ of distilled water and then a few drops of Universal Indicator solution. Compare the colour of the indicator with that obtained by adding a few drops of it to a solution of the original hydrated magnesium chloride in distilled water. Attempt to write an equation for the action of heat on hydrated magnesium chloride.

Repeat the experiment with the hydrated chlorides of calcium, strontium and barium. Record and comment on your findings.

3.10 Appearance and Solubility of Salts of the Group 2A Metals

(a) General appearance

Note the colours of a large range of salts of magnesium, calcium, strontium and barium, including the chromates and barium permanganate. Which salts are coloured and what ions are responsible for the colour? Which Group 2A metals form many deliquescent salts?

(b) Solubility of some Group 2A metal salts

(i) *Precipitation of the carbonates*, $M^{2+}CO_3^{2-}$

To about $3\,\text{cm}^3$ of 0·2M magnesium chloride solution add about $3\,\text{cm}^3$ of 0·2M sodium carbonate solution and note whether any precipitate of magnesium carbonate is formed.

Repeat the experiment using 0·2M solutions of calcium, strontium and barium chlorides in place of the magnesium chloride solution and record and comment on your observations.

(ii) *Precipitation of the chromates*, $M^{2+}CrO_4^{2-}$

Proceed as above but replace the 0·2M solution of sodium carbonate by a 0·2M solution of potassium chromate. Record and comment on your observations.

(iii) *Precipitation of the phosphates*, $M^{2+}HPO_4^{2-}$

Proceed as above but replace the 0·2M solution of potassium chromate by a 0·2M solution of disodium hydrogen orthophosphate, $(Na^+)_2HPO_4^{2-}$. Record and comment on your observations.

(iv) *Precipitation of the sulphates*, $M^{2+}SO_4^{2-}$

Proceed as above but replace the 0·2M solution of disodium hydrogen orthophosphate by 0·2M sulphuric acid. Record and comment on your results.

You should find that two sulphates do not precipitate under these conditions, so in these two cases repeat the experiment with 0·4M solutions of the Group 2A chlorides and 0·4M sulphuric acid and leave the solutions to stand for some time. Which sulphate precipitates on standing under these conditions?

Can you arrange the four Group 2A sulphates in order of increasing solubility?

3.11 The Diagonal Relationship of Lithium and Magnesium

In general the differences between the first and second members of a particular Periodic Group are more pronounced than those between the second and third. Thus the chemical differences between lithium and sodium are more pronounced than those between sodium and potassium. In many aspects of its chemistry, lithium shows similarities with magnesium. This is due to the fact that lithium and magnesium have similar electropositivities. Since electropositivity increases from top to bottom in any Periodic Group and decreases from left to right across a particular Period, the increase in electropositivity in going down one place in the Periodic Table (magnesium is one place lower down than lithium) is compensated for by the decrease which occurs in moving across a period from left to right (magnesium is one place to the right of lithium).

Look at your results for the action of heat on lithium nitrate, sodium nitrate and magnesium nitrate. In what way does the action of heat on sodium nitrate differ from the action of heat on lithium and magnesium nitrates?

In what way does the addition of methanol to anhydrous lithium chloride differ from the addition of methanol to sodium chloride? How was this result explained? Unfortunately the experiment cannot be carried out with anhydrous magnesium chloride, since it is purchased as a hydrated salt but the action of methanol on magnesium ethanoate and on calcium ethanoate can be compared.

Action of methanol on magnesium ethanoate and calcium ethanoate

To about 3 cm³ of methanol in a test-tube add a very small portion of magnesium ethanoate a little at a time with shaking. What do you observe?

Repeat the experiment with calcium ethanoate. Is there any difference and if so why?

3.12 Summary

The chemistry of beryllium is not considered in this book and the following summary excludes a consideration of beryllium and its compounds.

   (a) The Group 2A metals are reactive (the reactivity increases from top to bottom in the Group) but the reactivity is less than that of comparable members of Group 1A. The metals are extracted by electrolysis of their fused halides.

(b) The metals react strongly with non-metals such as chlorine and oxygen to give white ionic compounds. Magnesium decomposes steam at about 500K to give the oxide and hydrogen, while the other metals decompose water with the formation of the metallic hydroxide and hydrogen. All the metals react violently with hydrochloric acid.

(c) Magnesium and calcium form the normal oxide, e.g. $Ca^{2+}O^{2-}$, while the other members of the Group form the peroxide in addition, e.g. $Ba^{2+}O^{2-}$ and $Ba^{2+}O_2^{2-}$. The oxides and peroxides produce alkaline solutions with water (slight in the case of magnesium oxide).

(d) The solubilities of the hydroxides increase in the order magnesium hydroxide, calcium hydroxide etc. They are alkaline in solution.

(e) The hydrides are ionic (except magnesium hydride) and react vigorously with water to give alkaline solutions and hydrogen.

(f) Ionic nitrides can be formed by direct combination and these react with water to give alkaline solutions and ammonia.

(g) Except for magnesium, compounds of the Group 2A metals give coloured flame tests—calcium (yellowish-red), strontium (crimson) and barium (pale green).

(h) The thermal stabilities of salts containing a common anion increase from top to bottom in the Periodic Group, but comparable compounds of the Group 1A metals are more thermally stable.

(i) The majority of the salts of the Group 2A metals are white unless the anion happens to be coloured. For example, the chromates are yellow and the permanganates are purple. They form several insoluble salts, for example, carbonates and hydrogen phosphates. The solubilities of the sulphates decrease down the Group, for example, magnesium sulphate is quite soluble in water but barium sulphate is practically insoluble.

(j) Like lithium in Group 1A, magnesium displays appreciable covalent character in many of its compounds; for example, magnesium ethanoate is appreciably soluble in methanol (an indication of covalent character) whereas calcium ethanoate is not.

4

GROUP 3B BORON, ALUMINIUM, GALLIUM, INDIUM AND THALLIUM

4.1 Some Physical Data of Group 3B Elements

	Atomic number	Electronic configuration	Ionisation energy/kJ mol^{-1} First	Second	Third	Density/ g cm^{-3}	M.p./ K	B.p./ K	Atomic Radius/ nm	Ionic Radius/ nm M^{3+}
B	5	2.3 $1s^22s^22p^1$	800		3659	2·53	2300	4200	0·080	0·020 Estimated value
Al	13	2.8.3 ...$2s^22p^63s^23p^1$	578	1816	2744	2·70	932	2720	0·125	0·050
Ga	31	2.8.18.3 ...$3s^23p^63d^{10}4s^24p^1$	579	1979	2952	5·90	303	2510	0·125	0·062
In	49	2.8.18.18.3 ...$4s^24p^64d^{10}5s^25p^1$	558	1817	2693	7·28	429	2320	0·150	0·081
Tl	81	2.8.18.32.18.3 ...$5s^25p^65d^{10}6s^26p^1$	589	1970	2866	11·85	577	1740	0·155	0·095

4.2 Some General Remarks about Group 3B

All these elements exhibit a group valency of three, but because of the very large input of energy that is necessary to form the 3-valent ions—the sum of the first three ionisation energies—their compounds when anhydrous are either essentially covalent or contain an appreciable amount of covalent character. Boron never forms a B^{3+} ion since the enormous amount of energy required to remove three electrons from a small atom cannot be repaid with the formation of a stable crystal lattice, even with the most electronegative fluorine atom. Gallium and indium both form a chloride with the empirical formula XCl_2 in which they *appear* to be 2-valent. Physical evidence, however, shows that the correct formulation of these chlorides is $X^I(X^{III}Cl_4)$ in which both gallium and indium exhibit valencies of one and three. The tendency towards univalency is even more pronounced in the case of thallium and is due to the reluctance of the two *s* electrons of the heavier elements to be transferred, or to participate in covalent bond formation. This is an example of the inert pair effect, which is exhibited by other elements in Groups 4B, 5B and 6B.

Some 1-valent compounds of thallium resemble corresponding compounds of the alkali metals; thus the oxide, Tl_2O, and the hydroxide, $TlOH$, are strong bases. Other compounds are similar to those of silver, e.g. thallium(I) chloride like silver chloride, is only sparingly soluble in water and is sensitive to light.

The electronic configurations of the boron and aluminium atoms are similar in as much as the penultimate shell has a noble gas configuration, whereas the penultimate shell of the gallium, indium and thallium atoms contain eighteen electrons.

Boron, which is non-metallic, and aluminium which is clearly metallic are considered separately. No experimental work on gallium, indium and thallium is included in this chapter.

4.3 Extraction of the Group 3B Elements

Boron can be obtained as an amorphous brown powder by treating borax (sodium tetraborate), $(Na^+)_2B_4O_7^{2-}$, with hydrochloric acid, igniting the boric acid, H_3BO_3, to give the oxide, B_2O_3, and finally reducing the latter with magnesium at high temperature:

$$B_2O_3 + 3Mg \rightarrow 2B + 3Mg^{2+}O^{2-}$$

Aluminium is obtained by the electrolysis of carefully purified aluminium oxide, $(Al^{3+})_2(O^{2-})_3$, dissolved in molten cryolite, $(Na^+)_3AlF_6^{3-}$, at about 1200K. The latter chemical acts as a solvent for the aluminium oxide.

Gallium, indium and thallium are produced by the electrolysis of aqueous solutions of their salts. They are soft, white and fairly reactive metals.

4.4 The Oxide and Hydroxide of Boron

(a) Action of water on boron oxide, B_2O_3

To about $2 cm^3$ of distilled water in a test-tube add about $0.3 g$ of boron oxide, B_2O_3. Now add 2 drops of Universal Indicator solution and note the result. What type of oxide is boron oxide?

(b) Formation of boron oxide by the action of heat on boric acid

Determine the mass of a crucible and then place in it about 1 g of orthoboric acid, H_3BO_3 (or $B(OH)_3$), and determine the mass accurately. Heat the crucible in a pipeclay triangle, gently at first and then more strongly, until no further change takes place. Note your observations during the heating and the appearance of the final product, and then determine the mass of the crucible and contents after allowing the crucible to cool to room temperature. Heat to constant mass.

Using your results and having consulted a table of relative atomic masses, determine whether your results are in reasonable agreement with the equation:

$$2H_3BO_3 \rightarrow B_2O_3 + 3H_2O$$

(c) Formation of orthoboric acid, H_3BO_3

To about 5 cm³ of distilled water in a test-tube add about 1 g of sodium tetraborate, $(Na^+)_2B_2O_7^{2-} \cdot 10H_2O$, and then 3 drops of phenolphthalein. What type of reaction does this salt undergo in aqueous solution? Now add concentrated hydrochloric acid drop by drop until the phenolphthalein is just colourless. Allow the mixture to stand and note the formation of a white precipitate of orthoboric acid.

Orthoboric acid is a giant molecule and H_3BO_3 is simply its empirical formula. Look up in an inorganic chemistry text book the structure of this acid which involves hydrogen bonding.

Orthoboric acid is a weak monobasic acid which reacts with water in a rather unusual way:

$$B(OH)_3 + 2H_2O \rightarrow B(OH)_4^- + H_3O^+$$

i.e. it removes OH^- from water molecules leaving H_3O^+. Given that the boron atom has three electrons in its outer shell and that it forms three single covalent bonds with the oxygen atoms of the OH groups, can you explain why it functions in this manner?

(d) Formation of methyl orthoborate, $(CH_3)_2BO_3$—a volatile ester

Place about 0.1 g of sodium tetraborate in a test tube and then add about 10 drops of methanol and 10 drops of concentrated sulphuric acid. Warm gently and ignite the volatile ester which burns with a characteristic flame.

The sulphuric acid first produces orthoboric acid and then acts as a dehydrating agent:

$$3CH_3OH + B(OH)_3 \rightarrow (CH_3)_3BO_3 + 3H_2O$$

(e) Formation of transition metal borate glasses

Make a small loop in the end of a platinum wire and then heat it in the Bunsen flame until it is red hot. Quickly dip the loop into some powdered sodium tetraborate and reheat until a bead of glass-like material forms in the loop. This glass is a mixture of sodium metaborate and boron oxide:

$$(Na^+)_2B_4O_7^{2-} \rightarrow 2Na^+BO_2^- + B_2O_3$$

Moisten the bead with a little water and then dip it into a minute amount of nickel sulphate (too much solid will spoil the experiment) and then heat in the Bunsen flame. Note the colour of the bead produced which is nickel metaborate. Repeat with minute amounts of the sulphates of chromium(III), cobalt(II) and manganese(II) and record the colours of the beads.

The beads can readily be removed from the platinum wire by heating followed by sudden jerking of the wire.

Consult an inorganic text book for an explanation of the glass-like nature of sodium metaborate and boron oxide (prepared as above and under Section 4.4(b)) and of the transition metal metaborates.

4.5 Some Reactions of Aluminium

Aluminium combines directly with oxygen, sulphur, nitrogen and the halogens when heated to a sufficiently high temperature. The oxide and fluoride are essentially ionic, the rest are predominantly covalent and macromolecular.

(a) Formation of aluminium hydroxide, Al(OH)$_3$, from the metal

Pure aluminium is a reactive metal but ordinarily there is a very thin oxide film on its surface which protects the underlying metal from atmospheric attack. Once this oxide film is removed the reactive nature of the metal can be demonstrated.

Place a small piece of aluminium foil in a test-tube and cover it with an aqueous solution of mercury(II) chloride (**caution: mercury compounds are very poisonous**). Leave for a few minutes then wash the aluminium with distilled water. Leave the aluminium foil on a gauze and examine it periodically. Touch it gingerly and note whether it becomes warm. You should notice the formation of a moss-like growth of aluminium hydroxide on the surface, which is the result of attack by the moisture in the air.

Treat another piece of aluminium foil in a similar manner but, once the moss-like growth of aluminium hydroxide begins to appear, place the metal in a test-tube and cover it with distilled water. Shake and examine the metal over a period of a few minutes. Can you detect the evolution of any gas (hydrogen)? Compare this result with that obtained by adding distilled water to a piece of untreated aluminium foil.

(b) Formation of aluminium iodide, AlI$_3$ (Demonstration)

Place about 10 g of iodine in a dry mortar and mix it thoroughly with about an equal volume of aluminium powder. Place the mixture in a **fume cupboard** and then add a few drops of water and stand clear. If there is no action, add a little more water until a reaction occurs. Describe what you observe.

In this reaction the water is acting as a catalyst.

(c) Action of acids on aluminium

(i) *Action of hydrochloric acid on the metal*

Add about 5 cm^3 of approximately 2M hydrochloric acid to 0·5 g of aluminium (foil or shavings) in a test-tube and look for signs of evolution of gas. Warm if necessary. Can you detect any hydrogen? Repeat the experiment but this time use concentrated hydrochloric acid.

(ii) *Action of concentrated nitric acid on the metal*

Add about 5 cm^3 of concentrated nitric acid to 0.5 g of aluminium foil in a test-tube and leave for some time. Compare the result with the action of concentrated nitric acid on copper turnings, a metal which is generally much less reactive than aluminium (**caution: do this experiment in a fume cupboard**). Can you suggest any reason for the difference in behaviour? Consult a text book for a possible explanation.

(d) **Action of sodium hydroxide solution on the metal**

Add about 5 cm^3 of 2M sodium hydroxide solution to about 0.5 g of aluminium (the powder is best). Place the test-tube in a rack, since the reaction may become quite violent and produce much frothing, particularly if aluminium powder is used. If there is no action after a time, warm gently and test for the evolution of hydrogen:

$$2Al + 2OH^- + 6H_2O \rightarrow 2Al(OH)_4^- + 3H_2$$

(e) **The reducing action of aluminium (Demonstration)**

Caution: this experiment should be carried out with great care on a large asbestos square behind a safety screen

Place a mixture of aluminium powder and freshly dried iron(III) oxide (previously shown to be non-magnetic) in the ratio of 1:3 by mass in a fireclay crucible, partially immersed in a tin can containing sand. Into a small hole in the mixture place a small quantity of magnesium powder and barium peroxide and then a fuse of magnesium ribbon. Initiate the reaction by lighting the magnesium fuse and notice that the reaction proceeds with tremendous liberation of heat and emission of light. The temperature attained is well in excess of the melting point of iron and the metal can be extracted as a solid lump and shown to be magnetic.

$$2Al + Fe_2O_3 \rightarrow Al_2O_3 + 2Fe$$

4.6 Aluminium Oxide, $(Al^{3+})_2(O^{2-})_3$

(a) **Action of acid and alkali on aluminium oxide**

Place about 1 g of aluminium nitrate in a test-tube and heat gently until no further change occurs:

$$4Al^{3+}(NO_3^-)_3 \cdot 9H_2O \rightarrow 2(Al^{3+})_2(O^{2-})_3 + 12NO_2 + 3O_2 + 36H_2O$$

Extract the residue of aluminium oxide, placing half of it into one test-tube and the rest into another. Add approximately M sulphuric acid to one portion and warm. Note whether the aluminium oxide dissolves, then write an equation for the reaction. To the other portion of aluminium oxide add approximately 2M sodium hydroxide solution and warm the mixture. Again note whether the aluminium oxide dissolves.

Oxides of metals that react with both acids and alkalis are called **amphoteric oxides**.

(b) Aluminium oxide as an absorbing stationary phase in column chromatography—the separation of permanganate and dichromate ions

Aluminium oxide that has been heated strongly resists attack by both acids and alkalis (it was for this reason that the aluminium oxide used in Experiment 4.6(a) above was obtained by gently heating aluminium nitrate). The inactive form of aluminium oxide is frequently used in chromatography work.

Make up the chromatography tube (see Fig. 4.1) by placing a small wad of glass wool (about 2 cm in length) in the bottom of the tube. Then two-thirds fill the tube with 0·5M nitric acid and add the aluminium oxide carefully, opening the tap and allowing the excess acid to run away. Continue in this way until the tube is half full of aluminium oxide. Make sure that at no time is the top of the aluminium oxide not covered with acid, since this causes uneven packing of the column.

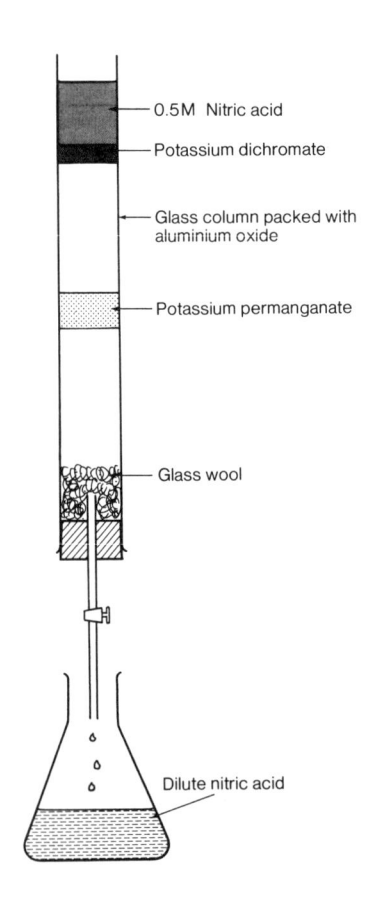

Fig. 4.1
Separation of permanganate and dichromate ions.

Carefully pipette on to the top of the column 10 cm³ of an aqueous mixture of potassium permanganate (potassium manganate(VII)) and potassium dichromate (0·02M with respect to potassium permanganate and 0·02M with respect to potassium dichromate). Open the tap at the bottom of the column and add 0·5M nitric acid so that the potassium permanganate passes down the column as a rather diffuse purple band. Continue adding the nitric acid and when the band nears the bottom of the column start to collect it in a conical flask. Remove the flask when all the potassium permanganate has passed through and leave it on one side.

Now change the eluting solution to M sulphuric acid and collect the potassium dichromate solution in a conical flask in a similar manner.

(i) *Estimation of the potassium permanganate solution*

Titrate the potassium permanganate solution in the conical flask with 0·1M ammonium iron(II) sulphate solution. Proceed slowly as the titre is very small, the end-point being when the purple colour of the permanganate ions has just been discharged and replaced by a slight tinge of green (iron(II) ions). Calculate the strength of the potassium permanganate solution (it should be 0·02M if the separation is quantitative).

$$MnO_4^- + 8H^+ + 5e^- \rightarrow Mn^{2+} + 4H_2O$$

$$Fe^{2+} \rightarrow Fe^{3+} + e^-$$

If the lower equation is multiplied by five and then the two equations added, the following equation is obtained:

$$MnO_4^- + 8H^+ + 5Fe^{2+} \rightarrow Mn^{2+} + 4H_2O + 5Fe^{3+}$$

or

$$MnO_4^- \equiv 5Fe^{2+}$$

(ii) *Estimation of the potassium dichromate solution*

Titrate the potassium dichromate solution in the conical flask in the following way. Add to it about 5 cm³ of 50% (V/V) orthophosphoric acid and 3 drops of barium diphenylamine sulphonate indicator. Carefully run into this mixture 0·1M ammonium iron(II) sulphate solution, proceeding slowly as the titre is again very small. As the titration proceeds, the solution in the flask becomes blue and the end-point, which is very sharp, is reached when this blue colour changes to green. Again, calculate the strength of the potassium dichromate solution (it should be 0·02M if the separation is quantitative).

$$Cr_2O_7^{2-} + 14H^+ + 6e^- \rightarrow 2Cr^{3+} + 7H_2O$$

$$Fe^{2+} \rightarrow Fe^{3+} + e^-$$

$$Cr_2O_7^{2-} \equiv 6Fe^{2+}$$

(c) Aluminium oxide as a dehydrating agent—formation of ethene from ethanol

Set up the apparatus as shown in Fig. 4.2. Pour some ethanol into the

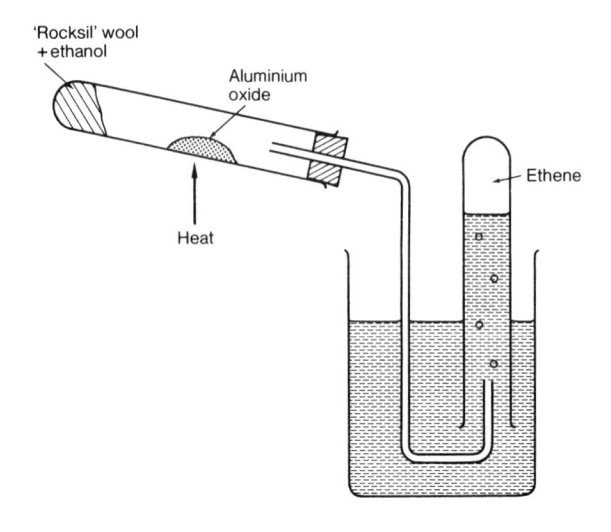

Fig. 4.2
Formation of ethene from ethanol.

test-tube to a depth of about 3 cm and then add sufficient 'Rocksil' wool to soak it up. Place about 1 g of aluminium oxide, in a small pile, half-way along the tube, and then insert a cork carrying a delivery tube. Heat the aluminium oxide with a small flame and collect several test-tubes of ethene by displacement of water.

$$CH_3-CH_2-OH \xrightarrow[600K]{Al_2O_3} H_2C=CH_2 + H_2O$$

Show that the ethene burns to give carbon dioxide (lime water turns milky). Also add some dilute bromine water to another test-tube containing ethene and show that it is decolorised.

4.7 Aluminium Hydroxide, Al(OH)$_3$

(a) Formation and amphoteric nature of aluminium hydroxide

Take about 2 cm^3 of approximately 0·1M aluminium sulphate solution in a test-tube and add to it drop by drop an approximately 2M solution of sodium hydroxide. Note the formation of any precipitate and whether it dissolves in an excess of sodium hydroxide solution.

Repeat the experiment, but stop the addition of sodium hydroxide solution once a precipitate appears. Now add some, approximately M sulphuric acid and note that the precipitate dissolves.

These results can be explained thus:

$$Al^{3+} + 3OH^- \longrightarrow Al(OH)_3$$

$$3OH^- + Al^{3+} \rightleftharpoons Al(OH)_3 + \text{Water} \rightleftharpoons Al(OH)_4^- + H_3O^+$$

$$\updownarrow 3H_3O^+ \qquad\qquad\qquad\qquad\qquad\qquad\qquad \updownarrow OH^-$$

$$6H_2O \qquad\qquad\qquad\qquad\qquad\qquad\qquad\qquad 2H_2O$$

See p. 43 for a fuller description of amphoteric behaviour.

(b) Adsorption of dyes by aluminium hydroxide

Place about 10 cm³ of a 0·1M solution of aluminium sulphate in a boiling-tube and then add 20 cm³ of approximately 2M aqueous ammonia to precipitate aluminium hydroxide. Now add about 5 cm³ of an approximately 1% aqueous solution of the dye alizarin red and stir the mixture well with a glass rod. Filter and note the colours of the aluminium hydroxide precipitate and the filtrate. What has happened?

4.8 Aluminium Chloride, Al₂Cl₆

Anhydrous aluminium chloride is a covalent solid having the formula $(AlCl_3)_n$. Relative molecular mass determinations in solution and in the vapour state show that under these conditions it exists as the dimer Al_2Cl_6. Consult an inorganic text book for an explanation of the bonding.

(a) Preparation of anhydrous aluminium chloride (Demonstration)

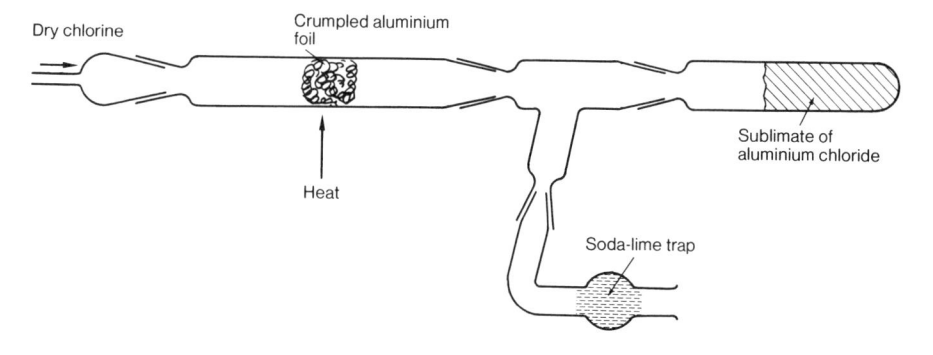

Fig. 4.3
Preparation of anhydrous aluminium chloride.

Set up the apparatus as shown in Fig. 4.3, in a fume cupboard, placing about 0·25 g of crumpled-up aluminium foil in the combustion tube. Heat the aluminium foil gently and then pass a stream of dry chlorine over it (preferably from a cylinder). Continue heating until most of the aluminium has reacted and the aluminium chloride produced has sublimed into the receiver. Turn off the supply of chlorine and stopper the receiver containing the anhydrous product.

(b) Some experiments with anhydrous aluminium chloride

(i) *Action of heat on anhydrous aluminium chloride*

Place about 0·2 g of the solid in a test-tube and heat it very gently. What do you observe?

(ii) *Action of water on anhydrous aluminium chloride*

Place about 0·2 g of the solid in a test-tube and then add distilled water, drop by drop. What do you observe? See if the tube becomes warm, and then add a few drops of Universal Indicator solution. What has happened and can you explain why?

(iii) *Action of ethanol on anhydrous aluminium chloride*

Examine the effect of ethanol on a small sample of anhydrous aluminium chloride contained in a test-tube. What do you observe? Does the tube become warm? Add a piece of blue litmus paper and notice if there is any effect on it. Now add distilled water gradually and notice if anything happens to the litmus paper. Do you observe any other change, and, if so, attempt to explain it?

4.9 A Comparison of the Acid-Base Properties of the Hydrated Aluminium and Magnesium Ions

To about 5 cm^3 of an aqueous solution of aluminium sulphate which is about 0·1M with respect to the Al^{3+} ion add about 3 drops of Universal Indicator solution and note the result. Repeat the experiment exactly, but this time use an aqueous solution of magnesium sulphate which is about 0·1M with respect to the Mg^{2+} ion.

In solution, the aluminium and magnesium ions are thought to be hydrated by six water molecules. There is the possibility of these hydrated ions reacting with other water molecules thus:

$$[Al(H_2O)_6]^{3+} + H_2O \rightleftharpoons [Al(H_2O)_5(OH)]^{2+} + H_3O^+ \qquad \text{etc.}$$

$$[Mg(H_2O)_6]^{2+} + H_2O \rightleftharpoons [Mg(H_2O)_5(OH)]^{+} + H_3O^+ \qquad \text{etc.}$$

In the above two reactions a water molecule abstracts a proton, H$^+$, from the hydrated cation, i.e. the water molecule is said to act as a Brønsted–Lowry base and the hydrated cation as a Brønsted–Lowry acid (consult an inorganic text book for a fuller discussion of Brønsted–Lowry acids and bases).

From your results decide which is a better Brønsted–Lowry acid, hydrated Al^{3+} or hydrated Mg^{2+}. Attempt to explain your results. How would you expect comparable strength solutions, containing, respectively, hydrated Na^+, Be^{2+}, Fe^{2+} and Fe^{3+}, to behave?

4.10 Aluminium Potassium Sulphate, $K^+Al^{3+}(SO_4^{2-})_2 \cdot 12H_2O$,—a Double Salt

When a solution containing potassium, aluminium and sulphate ions is allowed to crystallise, transparent octahedral crystals of aluminium potassium sulphate are obtained. This is called a double salt since it gives the characteristic reactions of its constituent ions in solution.

Preparation of aluminium potassium sulphate

Take 0·02 mole of potassium sulphate, $(K^+)_2 SO_4^{2-}$ (relative formula mass 174) and make a fairly concentrated solution in distilled water. Now make a similarly concentrated solution of aluminium sulphate by dissolving 0·02 mole of it, $(Al^{3+})_2(SO_4^{2-})_3 \cdot 18H_2O$ (relative formula mass 666) in some distilled water. Mix the two solutions, pour the resulting mixture into an evaporating dish and allow to crystallise. Wash the resulting crystals with a little water and dry with filter paper.

Pick out a reasonably well-shaped crystal and attempt to grow a larger one from a saturated solution of aluminium potassium sulphate.

Show that this double salt gives the test for the Al^{3+} ion (reaction of an aqueous solution with sodium hydroxide solution p. 46).

4.11 Summary

(a) Boron is non-metallic and is extracted by reduction of boron oxide with magnesium at high temperature.
(b) There is an extensive chemistry of boron based on the formation of boron-oxygen bonds; for example, bonds of this type are present in boron oxide, orthoboric acid and the borates (which have varied structures).
(c) Boron oxide, B_2O_3, is covalent and macromolecular. It slowly reacts with water forming orthoboric acid, H_3BO_3, which is also covalent and macromolecular.
(d) Aluminium is extracted by electrolysis of aluminium oxide dissolved in molten cryolite, $(Na^+)_3 AlF_6^{3-}$, at a temperature of about 1200K. It is a fairly reactive metal but ordinarily a protective oxide film on its surface protects the underlying surface from atmospheric attack. When the film is removed, by treatment with mercury(II) chloride

solution, for example, it will react with air and water at ordinary temperatures.

(e) Aluminium is attacked by concentrated hydrochloric acid with the evolution of hydrogen, but it is made passive by concentrated nitric acid. It reacts with sodium hydroxide solution forming an aluminate at the same time liberating hydrogen. The reaction between aluminium powder and many metallic oxides, for example iron(III) oxide, is highly exothermic.

(f) Aluminium oxide, $(Al^{3+})_2(O^{2-})_3$, is an amphoteric oxide. When it has been heated strongly it is chemically rather unreactive, and this inactive form of the oxide is used in column chromatography as the stationary phase and also as a catalyst in the dehydration of alcohols to alkenes, for example, ethanol to ethene. Aluminium hydroxide, like aluminium oxide, is amphoteric, and will readily adsorb dyes from solution.

(g) Aluminium chloride, $(AlCl_3)_n$, is covalent and macromolecular. It may be made by direct combination. The chloride is soluble in ethanol (an indication of covalent character) and reacts with water exothermically to give the hydrated aluminium cation, $Al^{3+}(aq)$, which shows an acid reaction in solution.

(h) Aluminium forms double sulphates, e.g. $K^+Al^{3+}(SO_4^{2-})_2 \cdot 12H_2O$.

5

GROUP 4B CARBON, SILICON, GERMANIUM, TIN AND LEAD

5.1 Some Physical Data of Group 4B Elements

	Atomic number	Electronic configuration	Density/ g cm^{-3}	M.p./ K	B.p./ K	Atomic Radius/ ·nm	Ionic radius/nm M^{2+}	M^{4+}
C	6	2.4 $1s^22s^22p^2$	2·22 (graphite)	> 3700	4120 (Subl)	0·077		
Si	14	$2s^22p^63s^23p^2$	2·42	1683	2570	0·117		0·041
Ge	32	2.8.18.4 ...$3s^23p^63d^{10}4s^24p^2$	5·46	1210	3020	0·122	0·093	0·053
Sn	50	2.8.18.18.4 ...$4s^24p^64d^{10}5s^25p^2$	7·31 (white Sn)	505	2630	0·141	0·112	0·071
Pb	82	2.8.18.32.18.4 ...$5s^25p^65d^{10}6s^26p^2$	11·34	600	2020	0·154	0·120	0·084

5.2 Some General Remarks about Group 4B

All these elements exhibit a group valency of four, but because an enormous amount of energy is needed to remove four electrons from their atoms, they form compounds which are predominantly covalent. Similarly, the gain of four electrons to give the 4-valent anion is energetically impossible.

Germanium, tin and lead form 2-valent compounds in which the two s electrons are inert (the inert pair effect). The stability of this state relative to the 4-valent state increases steadily from germanium to lead, i.e. 2-valent germanium compounds tend to be strongly reducing and revert to the 4-valent state, while for lead this is the predominant valency state. 2-valent compounds of tin and lead are often predominantly ionic.

Carbon and silicon are non-metallic, germanium is a metalloid and tin and lead are weakly metallic. This trend is reflected in the properties and structures of the elements; thus diamond is a non-conductor of electricity (graphite, however, will conduct), silicon and germanium are semiconductors and adopt the diamond structure (each atom is surrounded tetrahedrally by four more, so that the final structure is a three dimensional macromolecule), whereas tin and lead are conductors. Tin exhibits enantiotropy, one form having the diamond structure and the other two having closer packed atoms, this being a feature of metals; lead, on the other hand, only exists in one form which is metallic.

The first member of this group, carbon, differs considerably in its chemistry from silicon, the latter element having much in common with boron, its diagonal neighbour in the periodic table. The chemistry of carbon is dominated by its tendency to form chains and rings of carbon atoms in which other atoms, particularly hydrogen, play an important part. The chemistry of silicon is completely different and here the important feature is the formation of silicon–oxygen bonds, which are present in the giant molecule silicon(IV) oxide, and in the polymeric silicate anions.

The experimental chemistries of carbon and silicon are considered side by side to emphasise differences, germanium is not dealt with in this chapter, and tin and lead are treated together.

5.3 Extraction of the Group 4B Elements

Carbon is mined as graphite and diamond, although graphite is also manufactured from coke, and synthetic diamonds are made for industrial applications (see an inorganic textbook for details).

Silicon is obtained from silicon(IV) oxide by reduction with carbon in an electric furnace:

$$SiO_2 + 2C \rightarrow Si + 2CO$$

The final stages in the extraction of germanium involve reduction of germanium(IV) oxide with either carbon or hydrogen:

$$GeO_2 + 2H_2 \rightarrow Ge + 2H_2O$$

Tin is obtained by reduction of tin(IV) oxide with carbon and one process for extracting lead involves roasting lead(II) sulphide in air to give lead(II) oxide, which is then reduced to the metal with carbon:

$$SnO_2 + 2C \rightarrow Sn + 2CO$$
$$2PbS + 3O_2 \rightarrow 2PbO + 2SO_2$$
$$PbO + C \rightarrow Pb + CO$$

CARBON AND SILICON

5.4 Some Reactions of Carbon and Silicon

(a) Reaction of carbon and silicon with oxygen

Fill a boiling-tube with oxygen from a cylinder and cork it until it is used. Then place a small amount of powdered charcoal in a combustion spoon and heat it strongly until it is red hot. Place the combustion spoon in the boiling-tube full of oxygen and note what happens. When the reaction is over, pour a small amount of lime water into the boiling-tube, shake, and note the result. Write equations for the reactions involved.

Place a small amount of powdered silicon on a piece of porcelain and support this on a pipeclay triangle resting on a tripod. Heat the porcelain very strongly from below and then heat the silicon from above with a second Bunsen burner. If you manage to get a reaction to occur you will be left with a white solid which is silicon(IV) oxide (empirical formula SiO_2):

$$Si + O_2 \rightarrow SiO_2$$

Carbon dioxide and silicon (IV) oxide are both covalent compounds but the former is a gas and the latter a high melting point solid; clearly they have completely different structures. Attempt to write down the molecular structures of these two oxides before consulting a textbook.

(b) Action of steam on carbon and silicon (Demonstration)

Carry out Experiment 1.3(c), p. 11, if it has not been done previously. You should find that both carbon and silicon react with steam at a high temperature to give hydrogen:

$$C + H_2O \rightarrow CO + H_2$$
$$Si + 2H_2O \rightarrow SiO_2 + 2H_2$$

(c) Action of sodium hydroxide solution on carbon and silicon

Treat small samples of powdered charcoal and powdered silicon with approximately 2M sodium hydroxide solution in separate test-tubes. Look for signs of evolution of gas and warm gently if necessary. If any gas is evolved test for hydrogen. Do carbon and silicon react in the same way with sodium hydroxide solution?

(d) Reducing action of charcoal

Mix a small quantity of lead(II) oxide with about twice its volume of powdered wood charcoal and heat the mixture strongly on a piece of asbestos paper (about 3 cm × 3 cm). Globules of lead are soon formed which are soft enough to mark paper.

$$PbO + C \rightarrow Pb + CO$$

5.5 The Oxides of Carbon and Silicon

Carbon forms at least three oxides, and claims have been made that a further two exist. The two most common oxides of this element are carbon dioxide, CO_2, and carbon monoxide, CO. Silicon forms only one oxide, silicon(IV) oxide, SiO_2, which is a high melting point solid and thus macromolecular; it does, however, exist in a number of different crystalline forms, for example, quartz, tridymite and cristobalite.

(a) Formation and test for carbon dioxide

(i) *Action of heat on some carbonates*

Heat about 0·5 g of copper(II) carbonate in an ignition tube, gently at first and then more strongly. Pass any gas into a test-tube containing about 2 cm^3 of lime water. With the thumb over the test-tube, shake the lime water, and note whether it turns milky.

Repeat the experiment with zinc carbonate and dry 'reagent' grade sodium carbonate. Do these three carbonates evolve carbon dioxide with equal ease?

(ii) *Action of heat on some hydrogen carbonates*

Only the hydrogen carbonates of the Group 1A metals are thermally stable enough to exist as solids at room temperature.

Heat about 0·5 g of sodium hydrogen carbonate, $Na^+HCO_3^-$, in an ignition tube and test for carbon dioxide as above. What else do you observe?

Repeat with potassium hydrogen carbonate, $K^+HCO_3^-$.

(iii) *Action of hydrochloric acid on some carbonates and hydrogen carbonates*

Place about 0·5 g of each of the above carbonates and hydrogen carbonates in separate test-tubes and to each in turn add about 3 cm^3 of approximately 2M hydrochloric acid. Test for carbon dioxide in the usual manner. Write equations for the action of dilute hydrochloric acid on sodium carbonate and on sodium hydrogen carbonate.

(b) Reduction of carbon dioxide with magnesium (Demonstration)

Coil magnesium ribbon round a pencil (to form a spiral), remove the pencil, light the magnesium and lower it into a Pyrex flask filled with carbon dioxide. Allow the reaction to finish, and the flask to cool down, and then add some approximately 2M hydrochloric acid to dissolve the magnesium oxide. Carbon is left in the flask.

Which of the two elements magnesium or carbon is a better reducing agent under these conditions? How does your answer compare with the relative reducing properties of lead and carbon? (See Experiment 5.4(d).)

(c) **Formation and test for carbon monoxide**

Place about 0·5 g of sodium methanoate, $HCOO^-Na^+$, in a test-tube and then add about $2\,cm^3$ of concentrated sulphuric acid. Warm very gently (**care**) and ignite the carbon monoxide with a lighted splint.

$$HCOO^-Na^+ + H_2SO_4 \rightarrow HCOOH + Na^+HSO_4^-$$

$$HCOOH \rightarrow CO + H_2O$$

(d) **Reducing action of carbon monoxide (Demonstration)**

Set up the apparatus shown in Fig. 5.1 placing about $50\,cm^3$ of concentrated sulphuric acid in the $250\,cm^3$ flask. Slowly add methanoic acid from the tap funnel and note that enough heat is generated to make the reaction take place. The rate of production of carbon monoxide can be controlled by the rate of addition of the methanoic acid. Allow the reaction to proceed for several minutes and the carbon monoxide to escape into a well-ventilated fume cupboard (**caution: carbon monoxide is an exceptionally poisonous gas**). Now begin to heat the combustion tube containing the lead(II) oxide and go on heating until it has been converted into lead. What happens to the lime water once the lead(II) oxide is heated?

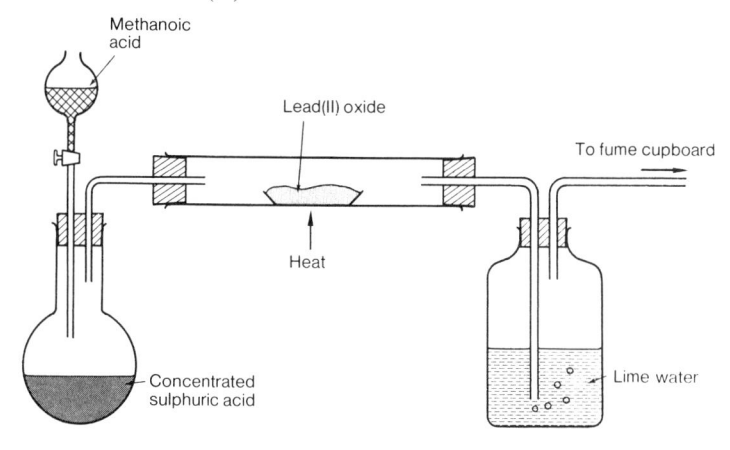

Fig. 5.1
Reduction of lead(II) oxide with carbon monoxide.

(e) **Absorption of carbon monoxide by an aqueous ammonia solution of copper(I) chloride**

Place about $50\,cm^3$ of an aqueous ammonia solution of copper(I) chloride in a shallow dish. Invert a test-tube containing carbon monoxide in this solution and clamp in an upright position. Leave for several hours and then examine.

Carbon monoxide forms an addition compound with the solution of copper(I) chloride:

$$CuCl + 2H_2O + CO \rightarrow CuCl\,.\,CO\,.\,2H_2O$$

(f) Formation of silicon(IV) oxide and reduction with magnesium

Silicon(IV) oxide may be obtained by heating powdered silicon with oxygen at about 700K. Like carbon dioxide, it is reduced by heating with magnesium but the reaction can be very violent and is best not attempted:

$$SiO_2 + 2Mg \rightarrow 2MgO + Si$$

5.6 The Acids Based on Carbon Monoxide, Carbon Dioxide and Silicon(IV) Oxide and their Salts

Carbon monoxide is not a true acidic oxide since it does not react with water to give an acid. However, it will react with fused sodium hydroxide under pressure to give sodium methanoate from which methanoic acid may be obtained by acidification with an aqueous solution of a stronger acid:

$$CO + Na^+OH^- \rightarrow HCOO^-Na^+$$

$$HCOO^-Na^+ + H_3O^+ \rightarrow HCOOH + H_2O + Na^+$$

Carbon dioxide reacts with water to give some carbonic acid which is only known in solution. Boiling the solution reverses the reaction below and carbon dioxide is evolved:

$$CO_2 + H_2O \rightleftharpoons H_2CO_3 \rightleftharpoons H^+ + HCO_3^- \rightleftharpoons 2H^+ + CO_3^{2-}$$

Despite the fact that carbon dioxide is known only in solution, solid carbonates and solid alkali metal hydrogen carbonates are well known.

Silicon(IV) oxide is not appreciably soluble in water but it does react with a hot concentrated solution of sodium hydroxide to give a number of sodium silicates, which contain polymeric silicate anions (see p. 59). Acidification of this solution produces polymeric silicic acid molecules. Silicate ions are therefore quite unlike the discrete carbonate ion.

(a) Methanoic acid and the action of heat on methanoates

The action of concentrated sulphuric acid on methanoic acid and sodium methanoate has previously been used to produce carbon monoxide (Experiment 5.5(c) and (d)). Methanoic acid is a colourless pungent liquid which is readily soluble in water.

(i) *Action of heat on sodium methanoate*, $HCOO^-Na^+$

Place about 0·5 g of sodium methanoate in an ignition tube and heat gently. Test for the evolution of hydrogen with a lighted splint. Now heat more strongly and see if you can detect any carbon monoxide and carbon dioxide. What is the dark residue left in the ignition tube?

(ii) *Action of heat on copper(II) methanoate*, $(HCOO^-)_2Cu^{2+}$

Place about 0·5 g of copper(II) methanoate in an ignition tube and heat gently at first and then more strongly. Test for the evolution of gases as in (i) above. What gases can you detect? What is the residue left in the ignition tube?

(b) Carbonic acid, carbonates and hydrogen carbonates

(i) *Solubility of carbon dioxide in water at room temperature*

Flush a 100 cm^3 syringe with carbon dioxide and then pass 50 cm^3 of the gas into it. Invert the syringe into a beaker of distilled water and suck about 20 cm^3 of water into it. Place the finger over the nozzle and shake. Replace the syringe in the water and note the volume of gas between the top of the water level and the piston and the volume of water in the syringe. Calculate the approximate volume of carbon dioxide that dissolves in 1 cm^3 of water at room temperature. Test the resulting solution to see how it affects blue litmus paper. Is the solution strongly or weakly acidic?

(ii) *Reaction of carbon dioxide with aqueous solutions of alkalis*

Place about 25 cm^3 of lime water in a boiling-tube and then pass carbon dioxide into it (from a Kipp's apparatus). Go on passing the carbon dioxide until no further change occurs, i.e. until the final mixture is a clear solution. Try to write two equations for the reactions that have taken place.

Boil the final clear solution and note what occurs. Try to write an equation for this reaction.

Carbon dioxide reacts with sodium hydroxide in a similar manner but, since sodium carbonate is soluble, the solution will at no time become cloudy.

(iii) *Alkaline natures of sodium carbonate and sodium hydrogen carbonate solutions*

Make up freshly prepared solutions of sodium carbonate and sodium hydrogen carbonate from analytical reagent grade chemicals which are approximately 1 mol dm^{-3} in strength. Use distilled water which has been boiled to expel any dissolved carbon dioxide. Into three test-tubes place about 10 cm^3 of the sodium carbonate solution and then add 2 drops of phenolphthalein to one tube, 2 drops of screened methyl orange to the second, and 2 drops of Universal Indicator solution to the third. Set up three similar tubes containing the sodium hydrogen carbonate solution and comment on the results. Consult a physical chemistry textbook to determine the pH range over which phenolphthalein and screened methyl orange change colour.

Take about 10 cm^3 of the sodium hydrogen carbonate solution and boil gently for a few minutes. Allow to cool and then add 2 drops of phenolphthalein. Explain the reaction that has taken place.

The carbonate ion, CO_3^{2-}, and the hydrogen carbonate ion, HCO_3^-, are Brønsted–Lowry bases and react with water abstracting a proton and leaving OH^-, e.g.

$$CO_3^{2-} + H_2O \rightleftharpoons HCO_3^- + OH^-$$

Write a similar equation for the reaction of the hydrogen carbonate ion with water. Which of the two anions would you expect to be a stronger Brønsted–Lowry base and why? Does this agree with your experimental results above?

(c) Silicic acid and silicates

(i) Formation of silicic acid gel

Dissolve some water glass (a viscous mixture which contains a number of sodium silicates including the orthosilicate anion SiO_4^{4-}) in several times its own volume of warm water and allow to cool. Titrate $25 cm^3$ of this solution with M hydrochloric acid using a few drops of phenolphthalein as indicator, the end-point being when the phenolphthalein turns colourless. The final product, silicic acid, should be a gel. Why does the solution of sodium silicate turn phenolphthalein pink?

The formation of silicic acid gel can be explained in the following manner. The orthosilicate anion, SiO_4^{4-}, is a strong base and reacts with hydrochloric acid thus:

$$SiO_4^{4-} + 4H^+ \rightarrow Si(OH)_4$$

followed by condensation with loss of water molecules:

$$nSi(OH)_4 \rightarrow H_{2n+2}Si_nO_{3n+1} + (n-1)H_2O$$

Silicic acid gel is a tangled mass of polysilicic acid molecules, the partial structure being indicated below:

```
      H     H     H     H     H     H
      |     |     |     |     |     |
      O     O     O     O     O     O
      |     |     |     |     |     |
---O—Si—O—Si—O—Si—O—Si—O—Si—O—Si---
      |     |     |     |     |     |
      O     O     O     O     O     O
      |     |     |     |     |     |
      H     H     H     H     H     H
```

The gel contains large amounts of water held in a cage-like structure of polysilicic acid. When it is heated this water is driven off and the polysilicic acid itself dehydrates, the material hardening and shrinking in the process. The end product is called silica gel—which of course is no longer a gel in the accepted sense of the word. It is a substance with an exceedingly large surface area and is used as a drying agent and as an inert supporting material for many finely divided catalysts.

(ii) Formation of silica gel from silicic acid gel

Place some of the silicic acid gel as prepared above, in an oven and heat it at a temperature of 100°C for several hours. Compare the texture of the silica gel produced with the silicic acid gel.

(iii) *The structure of silicates*

There is a bewildering variety of silicates, the basic features of their structures being readily understandable in terms of the linking together of tetrahedral SiO_4 units. Figure 5.2 shows a number of typical silicate anions, the negative charges on the silicate ions being exactly matched by the incorporation of a number of different cations in the final structures.

Examine Plate 1 (p. 60) which shows a piece of mica and a silicate rock containing some asbestos. What are the most noticeable features of these two

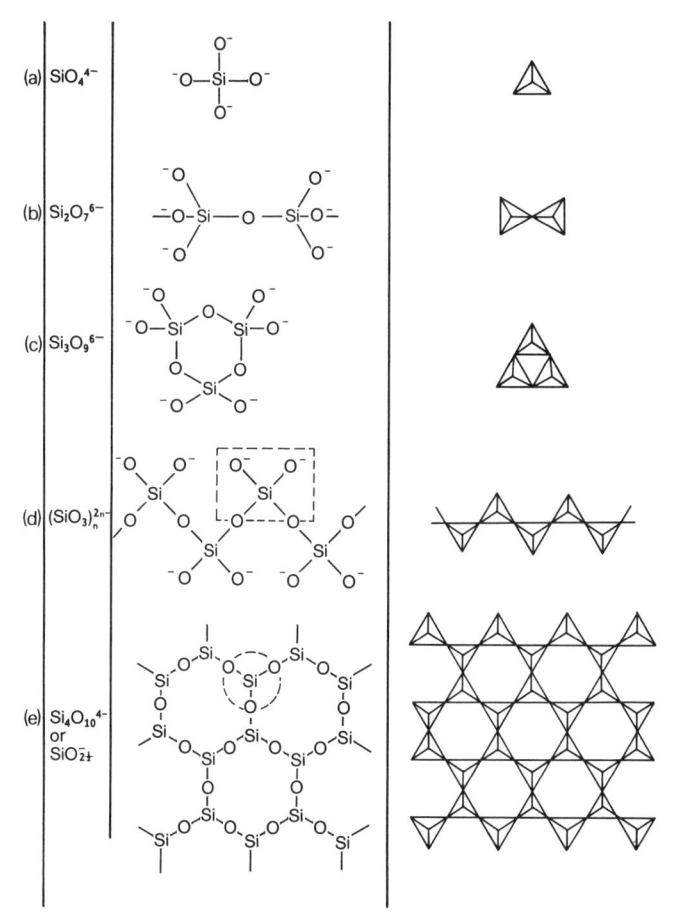

Fig. 5.2
The structures of silicate anions
(a) The structure of the orthosilicate anion.
(b) The structure of the pyrosilicate anion (one oxygen atom shared).
(c) The structure of a ring silicate anion (two oxygen atoms shared).
(d) The structure of an extended silicate anion (two oxygen atoms shared).
(e) The structure of an extended silicate anion (three oxygen atoms shared). In this structure the silicate atoms are joined to four oxygen atoms. The fourth oxygen atom carries a single negative charge and is above the plane of the diagram.

Plate 1
Photographs of mica (top) and a silicate rock (bottom) containing asbestos.

silicates? Can you relate them to any of the silicate structures shown in Fig 5.2?

5.7 The Silicones

These compounds are polymeric, the polymer chain containing alternately linked silicon and oxygen atoms. Alkyl or aryl groups, e.g. CH_3 or C_6H_5, are attached to the polymer backbone by means of covalent bonds to the silicon atom. A typical linear silicone has the formula:

$$R-\underset{R}{\overset{R}{Si}}-\left[O-\underset{R}{\overset{R}{Si}}\right]_n-O-\underset{R}{\overset{R}{Si}}-R$$

where R is an alkyl or aryl group.

Silicones are obtained by reacting an alkyl or aryl chloride with silicon in the presence of a copper catalyst at a temperature of about 600K. A mixture of alkyl or aryl chlorosilanes results:

$$6RCl + 3Si \xrightarrow{Cu} R_3SiCl + R_2SiCl_2 + RSiCl_3$$

These silane derivatives react with water to form 'hydroxides' which then condense with the elimination of water to form silicones (but R_3SiCl forms a dimer).

Preparation and properties of a silicone

Allow the **vapour** of dichlorodimethylsilane (**caution**) to pass through a burette using a filter pump. Also allow a filter paper to come into contact with the vapour.

The glass and paper surfaces contain adsorbed water molecules which react with the dichlorodimethylsilane to form a 'hydroxide' which then condenses with the elimination of water to form a silicone:

$$(CH_3)_2SiCl_2 + 2H_2O \rightarrow (CH_3)_2Si(OH)_2 + 2HCl$$

$$n(CH_3)_2Si(OH)_2 \rightarrow HO-\underset{CH_3}{\overset{CH_3}{Si}}-\left[O-\underset{CH_3}{\overset{CH_3}{Si}}\right]_{n-2}-O-\underset{CH_3}{\overset{CH_3}{Si}}-OH + (n-1)H_2O$$

Fill the treated burette with water and compare the meniscus with that of an untreated burette. What is the difference? Place the treated filter paper in a filter funnel and pour into it a mixture of water and kerosine. What is the result? Can you explain the water-repellent properties of a silicone?

Hopkins and Williams Ltd market silicones, one such preparation labelled MS 1007 being suitable for the following experiments.

Make up a 3% solution of the silicone fluid in tetrachloromethane and use it to wet thoroughly the inside of a clean test-tube. Drain off the liquid and allow the test-tube to dry in the air. Then bake it in an oven at about

100°C for about 2 hours. Pour water into the test-tube and examine the meniscus.

Treat a small portion of a thermally insulating brick with the silicone mixture used above, allow to dry in air and then bake in an oven as above. Place the treated brick in a bowl of water and note the result. Compare with the effect of water on an untreated brick.

5.8 The Hydrides of Carbon and Silicon

A limitless number of hydrides of carbon exist and owe their existence to the unique property of carbon atoms to catenate (link together) into stable structures. A few typical hydrocarbons are given below; a detailed study of these compounds, however, lies outside the scope of this chapter and is dealt with in organic chemistry textbooks.

	General Formula	Examples
The alkanes	C_nH_{2n+2}	CH_4, H_3C-CH_3, $H_3C-CH_2-CH_3$
The alkenes	C_nH_{2n}	$H_2C=CH_2$, $H_3C-CH=CH_2$
The alkynes	C_nH_{2n-2}	$H-C\equiv C-H$, $H_3C-C\equiv C-H$

Silicon does not catenate to anything like the same extent as carbon and, although a series of silanes analogous to the alkanes have been prepared with up to six silicon atoms in the chain, they are all spontaneously flammable and immediately hydrolysed by water.

There are no silicon analogues of the alkenes and alkynes showing the reluctance of the silicon atom to form multiple bonds.

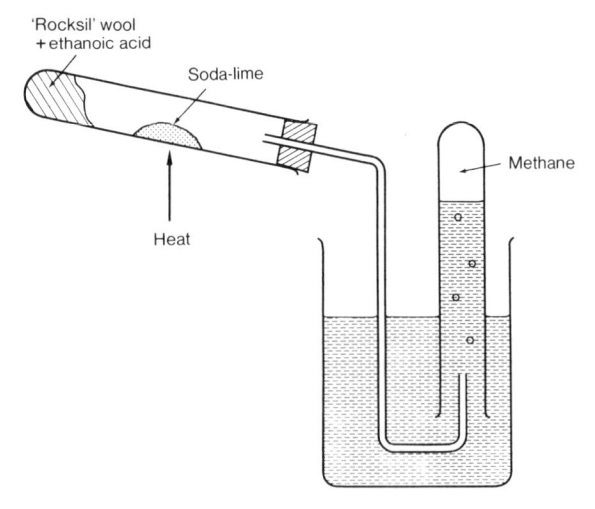

Fig. 5.3
The formation of methane.

(a) Formation of methane, CH_4

Pour glacial ethanoic acid into a test-tube to a depth of about 3 cm and then push some 'Rocksil' wool into the liquid to soak it up. Place about 1 g of soda-lime (a mixture of calcium hydroxide and sodium hydroxide) in a small pile, about half-way along the test-tube and then set up the apparatus as shown in Fig. 5.3. Heat the soda-lime with a small flame and then collect one or two test-tubes full of the gas over water.

Ignite the methane and note the colour of the flame. What are the products of combustion?

(b) Formation of a mixture of silanes including monosilane, SiH_4 (Demonstration)

Mix 2 parts by mass of magnesium powder with 1 part of dry amorphous silicon(IV) oxide and then place a small quantity of this mixture in a test-tube. Clamp the test-tube with the mouth of the tube pointing in a safe direction and then place a lighted Bunsen under the bottom of the tube, **standing clear**. A violent reaction should occur resulting in the formation of some magnesium silicide (but see Section 5.5(f) p. 56). When the tube has cooled down, stand it in an upright position and then add approximately 2M hydrochloric acid from a dropping tube. A fairly vigorous reaction should occur with the liberation of a mixture of silanes and some hydrogen (from any unreacted magnesium) which catches fire spontaneously.

$$Mg_2Si + H^+ \rightarrow Mg^{2+} + SiH_4 + H_3Si-SiH_3 \qquad \text{etc.}$$

Write an equation for the spontaneous reaction of monosilane with oxygen.

Consult an inorganic textbook for an explanation of the difference in stabilities of methane and monosilane with respect to oxygen.

5.9 The Chlorides of Carbon and Silicon

(a) Action of water on tetrachloromethane, CCl_4

Place about 1 cm^3 of tetrachloromethane in a test-tube and then add about the same volume of distilled water. Shake the mixture and allow to stand. Is there any reaction?

(b) Action of water on silicon tetrachloride, $SiCl_4$

Place about 1 cm^3 of distilled water in a test-tube and then add silicon tetrachloride drop by drop (**care**). Does the tube become warm? Now add a few drops of Universal Indicator solution and note what happens. Attempt to write down an equation for the reaction between silicon tetrachloride and water.

Can you explain the difference in stabilities of tetrachloromethane and silicon tetrachloride with respect to water? If not, consult an inorganic textbook.

TIN AND LEAD

5.10 Some Reactions of Tin and Lead

(a) Reaction of tin and lead with oxygen (air)

Place a piece of granulated tin on a piece of broken porcelain and heat over a pipe-clay triangle. Note any changes which occur. Occasionally stir the molten tin with a spatula to determine whether oxidation takes place throughout the metal. Examine after cooling.

Repeat the experiment with a piece of granulated lead and note any changes as above.

(b) Action of hydrochloric acid on tin and lead

Place a piece of granulated tin in a test-tube and add about $3\,cm^3$ of approximately 2M hydrochloric acid. Can you detect much reaction in the cold and on warming? Repeat the experiment with concentrated hydrochloric acid and warm if necessary. Can you detect any hydrogen? Collect over water if necessary (as in Fig. 1.4 p. 12).

Repeat the experiment with a piece of granulated lead and comment on your results.

(c) Action of dilute nitric acid on tin and lead

Place a piece of granulated tin in a test-tube and add about $3\,cm^3$ of approximately 2M nitric acid. Warm if there is no reaction in the cold. Comment on your results.

Repeat the experiment with a piece of granulated lead and comment on your findings.

Check the results of all these experiments by consulting an inorganic textbook.

5.11 The Oxides of Tin and Lead

Tin forms two oxides, tin(IV) oxide, which is a white solid and tin(II) oxide, which is a black solid. They have the respective empirical formulae SnO_2 and SnO. Lead forms three solid oxides, namely lead(IV) oxide, PbO_2, dilead(II) lead(IV) oxide, Pb_3O_4, and lead(II) oxide, PbO. Their colours are respectively dark brown, red and yellow, although slight variations are possible, depending upon their method of preparation.

(a) Formation of tin(IV) oxide, SnO_2

Determine the mass of a crucible and lid and then place in the crucible about 1 g of tin foil; determine the mass of the tin accurately. Place the crucible in a fume cupboard and add concentrated nitric acid drop by drop to the tin until no further action occurs. Place the lid on the crucible and heat fairly strongly over a pipe-clay triangle until no further evolution of brown gas occurs. Remove the lid and heat strongly for another minute or so. After cooling, determine the mass of crucible and lid plus the contents. Reheat to constant mass.

Using values for the relative atomic masses of tin and oxygen determine the number of moles of oxygen that combine with one mole of tin. Are your results in reasonable agreement with the formula SnO_2?

(b) Formation of tin(II) oxide, SnO

Place about 0·5 g of tin(II) oxalate in a test-tube and heat fairly strongly until no further change occurs. Carbon monoxide is produced and provides a reducing atmosphere preventing aerial oxidation to tin(IV) oxide:

$$\begin{matrix} COO^- \\ | \quad Sn^{2+} \\ COO^- \end{matrix} \rightarrow SnO + CO + CO_2$$

After the decomposition of the oxalate is complete, shake a little of the residue of tin(II) oxide on to an asbestos square and note what happens to the warm solid as it passes through the air. What has happened to the tin(II) oxide?

(c) Formation and some reactions of lead(IV) oxide, PbO_2

(i) *Formation of lead(IV) oxide by oxidation of an aqueous solution of a lead(II) salt*

Place about $2 cm^3$ of lead(II) nitrate solution in a test-tube and then add about the same volume of sodium chlorate(I) (sodium hypochlorite) solution. Warm gently until the mixture turns dark brown and then filter off the precipitate of lead(IV) oxide.

$$Pb^{2+} + H_2O + ClO^- \rightarrow PbO_2 + Cl^- + 2H^+$$

(ii) *Formation of lead(IV) oxide by the action of nitric acid on dilead(II) lead(IV) oxide (red lead)*

Place about 0·5 g of dilead(II) lead(IV) oxide (red lead) in a test-tube and then add about $3 cm^3$ of approximately 2M nitric acid. Warm gently and filter off the dark deposit of lead(IV) oxide. In this experiment dilead(II) lead(IV) oxide behaves as if it were a mixture of 2PbO, which reacts with the dilute nitric acid, and PbO_2 which does not:

$$Pb_3O_4 + 4HNO_3 \rightarrow 2Pb(NO_3)_2 + PbO_2 + 2H_2O$$

(iii) *Action of heat on lead(IV) oxide*

Heat about 0·5 g of lead(IV) oxide in an ignition tube and test for the evolution of a gas. What is left in the ignition tube?

(iv) *Action of concentrated hydrochloric acid on lead(IV) oxide*

Do this experiment in a fume cupboard. To about 0·5 g of lead(IV) oxide in a test-tube add about 2 cm^3 of concentrated hydrochloric acid and warm very gently if necessary. Hold a piece of moist red litmus paper near the mouth of the test-tube and comment on the result. What gas is evolved in this reaction? Attempt to write an equation for the reaction.

In view of what you have observed in the previous two experiments, which do you consider to be the most stable valency state of lead, Pb(II) or Pb(IV)? Is this also true for tin as far as the two oxides are concerned?

(d) Formation and action of heat on dilead(II) lead(IV) oxide, Pb$_3$O$_4$

Dilead(II) lead(IV) oxide can be obtained by heating lead(II) oxide in air at about 670K:

$$6PbO + O_2 \rightarrow 2Pb_3O_4$$

(i) *Action of heat on dilead(II) lead(IV) oxide*

Heat about 0·5 g of dilead(II) lead(IV) oxide in an ignition tube and test for the evolution of a gas. In view of the fact that this oxide behaves as if it were a mixture of 2PbO and PbO$_2$ what do you think is left in the ignition tube?

(ii) *Action of concentrated hydrochloric acid on dilead(II) lead(IV) oxide*

Do this experiment in a fume cupboard. To about 0·5 g of dilead(II) lead(IV) oxide in a test-tube add about 2 cm^3 of concentrated hydrochloric acid and warm very gently if necessary. Hold a piece of moist red litmus paper near the mouth of the tube and identify the gas evolved. Try to write an equation for this reaction.

5.12 The Hydroxides of Tin and Lead

Formation of tin(II) hydroxide, Sn(OH)$_2$, and lead(II) hydroxide, Pb(OH)$_2$, and their reactions with acid and alkali

Dissolve about 0·5 g of tin(II) chloride in the minimum volume of concentrated hydrochloric acid and then dilute with distilled water.

To two separate portions of about 3 cm^3 of the tin(II) chloride solution in two test-tubes add approximately 2M sodium hydroxide solution drop by drop until a fairly heavy precipitate is formed. Continue adding the sodium hydroxide solution to one of the test-tubes and stir with a glass rod; what happens and why? To the other test-tube add some concentrated hydrochloric acid and observe what happens. Try to write equations for the action of alkali and acid on tin(II) hydroxide.

Repeat the above experiments using a solution of lead(II) nitrate but substitute approximately 2M nitric acid for the hydrochloric acid. Are the results similar, and if so what does this mean about lead(II) hydroxide?

5.13 The Sulphides of Tin and lead

(a) Formation of tin(IV) sulphide, SnS_2, and tin(II) sulphide, SnS

Dissolve about 0·5 g of hydrated tin(IV) chloride, $SnCl_4 . 5H_2O$, in a little water, adding some concentrated hydrochloric acid, if necessary, to clear the solution. Pass a stream of hydrogen sulphide through about $3 cm^3$ of this solution (**caution: hydrogen sulphide is extremely unpleasant and poisonous**). If no precipitate appears, dilute the solution with a little distilled water. Note the colour of the precipitate. Now add approximately 2M sodium hydroxide solution to the resulting mixture and stir with a glass rod. What happens?

Since sulphur is in the same periodic group as oxygen it is reasonable to regard metallic sulphides as having some properties similar to oxides. What type of behaviour is the tin(IV) sulphide therefore displaying?

Repeat the above experiment using a solution of tin(II) chloride prepared as in Section 5.12. Is there any difference in behaviour of tin(II) sulphide as compared with tin(IV) sulphide or not?

(b) Formation of lead(II) sulphide, PbS

Pass a stream of hydrogen sulphide through about $3 cm^3$ of a solution of lead(II) nitrate and note the colour of the precipitate. Examine the effect of the addition of sodium hydroxide solution on the resulting mixture. Is the result similar to the action of sodium hydroxide solution on tin(II) sulphide? Can you explain this result in view of the relative positions of tin and lead in the periodic table?

5.14 The Hydrides of Tin and Lead

Unlike carbon and silicon, tin only forms one hydride stannane, SnH_4, which is a covalent gas. It may be obtained by reduction of tin(IV) chloride with lithium aluminium hydride at low temperature:

$$SnCl_4 + Li^+ AlH_4^- \rightarrow SnH_4 + Li^+ Cl^- + AlCl_3$$

It decomposes slowly at room temperature and rapidly on heating, but resists attack by 15% alkali.

Plumbane, PbH_4, is less well characterised but it has been shown to exist by treating an alloy of magnesium and lead (containing $^{212}_{82}Pb$ which is radioactive) with dilute acid. The detection of radioactivity in the gas phase shows that a volatile hydride of lead is formed.

5.15 The Chlorides of Tin and Lead

Tin forms two chlorides. Tin(IV) chloride, $SnCl_4$, is a colourless covalent liquid which reacts with a little water to give the hydrate, $SnCl_4 . 5H_2O$, which is ionic and presumably contains the $[Sn(H_2O)_4]^{4+}$ ion. Tin(II) chloride, $SnCl_2$, is a white covalent solid, which reacts with a little water to give the hydrate, $SnCl_2.2H_2O$. Lead also forms two chlorides, the 4-valent chloride, $PbCl_4$, being a colourless covalent liquid that readily decomposes into lead(II) chloride and chlorine. Lead(II) chloride is a white solid possessing some covalent character.

(a) Formation and some reactions of tin(IV) chloride, $SnCl_4$

(i) *Formation of tin(IV) chloride*

Set up, **in a fume cupboard**, the apparatus shown in Fig. 5.4 placing a layer of dry sand at the bottom of the 100 cm³ flask. On top of the sand place about 3–4 g of dry granulated tin and, after melting it with a small flame, pass chlorine over it, preferably from a cylinder. Continue to heat sufficiently strongly so that the tin(IV) chloride distils over into the receiver, fitted with a calcium chloride tube to prevent the entry of moisture. Redistil the product, collecting the liquid boiling over the range 112–115°C and immediately stopper the product.

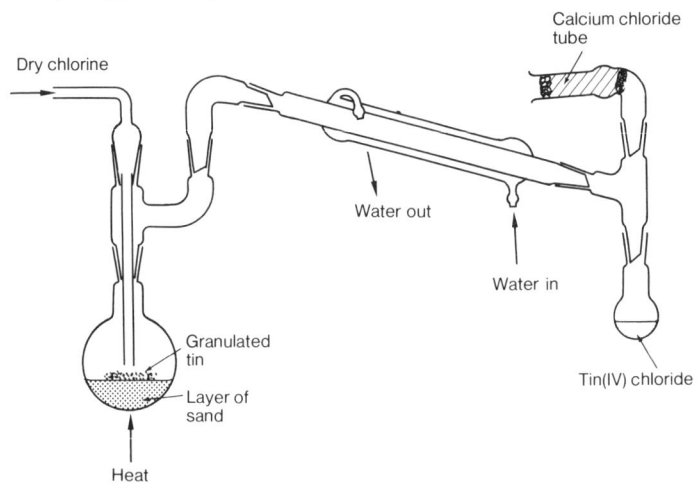

Fig. 5.4
The formation of tin(IV) chloride.

(ii) *Action of water on tin(IV) chloride*

In a fume cupboard unstopper the tube containing the tin(IV) chloride and notice what happens. Now place about 0·2 cm³ of the chloride in a test-tube and add 2 or 3 drops of distilled water. Shake the tube, note if it becomes warm, and then allow to cool. A white solid should be obtained which is hydrated tin(IV) chloride:

$$SnCl_4 + 5H_2O \rightarrow SnCl_4.5H_2O$$

Does the solid dissolve on the addition of more water?

(iii) *Action of tetrachloromethane on tin(IV) chloride*

Take about 0.2 cm^3 of tin(IV) chloride in a test-tube and add about 0.5 cm^3 of tetrachloromethane. Does the tube appear to get hot and do the two liquids mix? What does this experiment indicate to you about the bonding in tin(IV) chloride?

(b) Formation and some reactions of tin(II) chloride, $SnCl_2$

(i) *Formation of tin(II) chloride*

Place about 0.5 g of granulated tin in a test-tube and then add about 3 cm^3 of concentrated hydrochloric acid. Warm if necessary and then allow the reaction to proceed on its own for several minutes. Filter the solution of tin(II) chloride which is tested in the next experiment.

$$Sn + 2HCl \rightarrow SnCl_2 + H_2$$

(ii) *Reducing action of tin (II) chloride solution*

Dilute the tin(II) chloride obtained as above with an equal volume of distilled water and then divide it into two portions.

To one portion add iron(III) chloride solution drop by drop until the yellow colour persists after warming. Now add a little tin(II) chloride solution to discharge the yellow colour. Add a small amount of approximately 2M sodium hydroxide solution and observe the colour of any precipitate (other than a white one which may form and which is tin(II) hydroxide). Add some approximately 2M sodium hydroxide solution to separate portions of iron(III) chloride solution and iron(II) sulphate solution, and observe the colours of each precipitate. From these results deduce an equation for the action of tin(II) ions on iron(III) ions in aqueous solution.

To the second portion of tin(II) chloride solution add a small quantity of a dilute solution of mercury(II) chloride solution and note what happens. Now warm the mixture gently and observe any further change. In this experiment the tin(II) ions reduce the mercury(II) ions first to mercury(I) and then to metallic mercury:

$$2Hg^{2+} + Sn^{2+} \rightarrow [Hg-Hg]^{2+} + Sn^{4+}$$
$$[Hg-Hg]^{2+} + Sn^{2+} \rightarrow 2Hg + Sn^{4+}$$

(c) Formation and some reactions of lead(II) chloride, $PbCl_2$

(i) *Formation of lead(II) chloride*

Place about 3 cm^3 of $0.2M$ lead nitrate solution in a test-tube and add approximately 2M hydrochloric acid until no further precipitation occurs:

$$Pb^{2+} + 2Cl^- \rightarrow PbCl_2$$

Carry out the following experiments with lead(II) chloride prepared as above.

(ii) *Solubility of lead(II) chloride*

Dilute the mixture obtained as above with an equal volume of distilled water and warm in a beaker containing boiling water. What happens? Allow the mixture to cool and observe what happens. What can you say about the relative solubility of lead(II) chloride in hot water and in cold?

(iii) *Action of potassium iodide solution on lead(II) chloride solution*

Prepare some lead(II) chloride solution as in (i) above and filter. To the filtrate add about $3\,cm^3$ of $0.5M$ potassium iodide solution and observe what happens. What can you say about the relative solubilities of lead(II) chloride and lead(II) iodide?

(iv) *Action of concentrated hydrochloric acid on lead(II) chloride*

To about $3\,cm^3$ of $0.5M$ lead nitrate solution add concentrated hydrochloric acid until no further precipitation occurs. Now add more concentrated hydrochloric acid until the precipitate just dissolves. Add a fairly large volume of distilled water and observe what happens on standing. Explain what is happening in terms of the equilibrium reaction:

$$PbCl_2 + 2Cl^- \rightleftharpoons PbCl_4^{2-}$$

Would you expect lead(II) chloride solution to dissolve in the presence of other ionic chlorides? Would the solution need to be dilute or concentrated? Carry out further experiments to check your answers.

(d) Formation of ammonium hexachloroplumbate(IV), $(NH_4^+)_2PbCl_6^{2-}$, a derivative of lead(IV) chloride

Place about $0.5\,g$ of lead(IV) oxide in a test-tube and cool in ice water. To it add ice-cold concentrated hydrochloric acid and, after mixing, pour into it about $3\,cm^3$ of a fairly concentrated cold solution of ammonium chloride. Mix well and filter off the bright yellow precipitate of ammonium hexachloroplumbate(IV):

$$PbO_2 + 4HCl \rightarrow PbCl_4 + 2H_2O$$
$$2NH_4^+Cl^- + PbCl_4 \rightarrow (NH_4^+)_2PbCl_6^{2-}$$

5.16 The Oxysalts of Lead

Most of the lead(II) oxysalts are insoluble in water, notable exceptions being the nitrate, $Pb(NO_3)_2$, and the ethanoate, $Pb(CH_3COO)_2 \cdot 3H_2O$, which are freely soluble.

(a) Formation of lead(II) chromate, $PbCrO_4$

To about $2\,cm^3$ of a $0.5M$ solution of lead(II) nitrate add about $2\,cm^3$

of a 0·2M solution of potassium chromate and note the formation of a yellow precipitate of lead(II) chromate:

$$Pb^{2+} + CrO_4^{2-} \rightarrow PbCrO_4$$

(b) Formation of lead(II) sulphate, $PbSO_4$

To about $2\,cm^3$ of a 0·5M solution of lead(II) nitrate add about $2\,cm^3$ of approximately M sulphuric acid and note the formation of a white precipitate of lead(II) sulphate:

$$Pb^{2+} + SO_4^{2-} \rightarrow PbSO_4$$

The only stable lead(IV) oxysalt is the ethanoate, $Pb(CH_3COO)_4$, which as expected is an oxidising agent.

(c) Formation of lead(IV) ethanoate, $Pb(CH_3COO)_4$, and its reaction with water

Place about $75\,cm^3$ of glacial ethanoic acid in a small beaker and add to it about $25\,cm^3$ of ethanoic anhydride. Place this mixture **in a fume cupboard** and add small amounts of dilead(II) lead(IV) oxide, Pb_3O_4, to this mixture, warming gently to about 60°C. Stir the mixture well with a glass rod and go on adding small portions of dilead(II) lead(IV) oxide until there are signs of the formation of the dark coloured lead(IV) oxide. Filter the hot solution rapidly through a fluted paper into a conical flask. Immediately cork the conical flask and cool in ice-water to precipitate the lead(IV) ethanoate. Filter off the lead(IV) ethanoate using a Buchner funnel, dry the product between pads of filter paper and scrape the dried solid into a test-tube which should then be corked.

Place a small amount of the lead(IV) ethanoate in about $2\,cm^3$ of distilled water and observe what happens. What do you think is formed? Confirm your answer by experiment, preparing more of the product if need be.

5.17 Summary

(a) Carbon occurs as graphite and diamond, while silicon is extracted by reduction of silicon(IV) oxide with magnesium at high temperature.

(b) Carbon and silicon combine with oxygen on heating, the former more readily than the latter, giving respectively carbon dioxide and silicon(IV) oxide. Both elements reduce steam at high temperatures to hydrogen. The reducing action of carbon is employed industrially to obtain many metals from their oxides.

(c) Carbon dioxide, CO_2, and carbon monoxide, CO, are gaseous whereas silicon(IV) oxide is a high melting point solid whose structure is macromolecular, SiO_2 simply being its empirical formula.

(d) Carbon dioxide gives rise to carbonates and hydrogen carbonates and is a typical acidic oxide; carbon monoxide does not react with water but will react with fused alkalis to give methanoates.

Silicon(IV) oxide reacts with hot concentrated alkalis to give a number of silicates which are polymeric and numerous.
(e) There is an extensive chemistry of silicon based on silicon–oxygen bonds which are present in silicon(IV) oxide, the silicates and the silicones. The most important feature of carbon chemistry is the unique property of carbon atoms to catenate into stable structures, thus giving rise to the alkanes, the alkenes and the alkynes for example.
(f) Methane, CH_4, is the first member of the alkane series and burns to give carbon dioxide and water. Monosilane, SiH_4, is the silicon analogue but it is spontaneously inflammable as are the other silanes.
(g) Tetrachloromethane, CCl_4, is a covalent liquid which is immiscible with and unaffected by water. Silicon tetrachloride, $SiCl_4$, is also a covalent liquid but is readily hydrolysed by water.
(h) Tin and lead are obtained by reduction of the respective oxides SnO_2 and PbO (obtained from PbS by roasting in air) with carbon. Both metals are scarcely attacked by dilute hydrochloric acid but the concentrated acid attacks them. They are also attacked by dilute nitric acid.
(i) Tin forms two oxides, tin(IV) oxide, SnO_2, and tin(II) oxide, SnO; the latter is fairly readily oxidised to the former. Lead forms three oxides, lead(IV) oxide, PbO_2, dilead(II) lead(IV) oxide, Pb_3O_4, and lead(II) oxide, PbO, the first two evolving oxygen on heating, leaving lead(II) oxide.
(j) Tin(II) hydroxide and lead(II) hydroxide can both be precipitated as white solids by the addition of sodium hydroxide solution to solutions of tin(II) and lead(II) salts respectively. They dissolve in aqueous solutions of acids and strong alkalis and are thus amphoteric.
(k) Tin forms two sulphides, tin(IV) sulphide, SnS_2, and tin(II) sulphide, SnS, both of which dissolve in sodium hydroxide solution, i.e. they display 'acidic' properties. Lead only forms one sulphide, lead(II) sulphide, PbS, which is insoluble in sodium hydroxide solution, a property in keeping with the more 'metallic' character of this element.
(l) There are two chlorides of tin, tin(IV) chloride, $SnCl_4$, which is a covalent liquid readily hydolysed by water to give the $[Sn(H_2O)_4]^{4+}$ ion, and tin(II) chloride, $SnCl_2$, a white covalent solid which gives an ionic hydrate with water. In solution, tin(II) ions, Sn^{2+}, are readily oxidised to tin(IV) ions, Sn^{4+}, e.g. with Fe^{3+} ions. Although lead(IV) chloride, $PbCl_4$, exists, it readily changes into lead(II) chloride and chlorine, but a derivative of lead(IV) chloride—ammonium hexachloroplumbate(IV), $(NH_4^+)_2PbCl_6^{2-}$—may readily be prepared. Lead(II) chloride is a white solid which is much more soluble in hot water than in cold.
(m) Lead(II) iodide is more insoluble in water than lead(II) chloride. The former is a yellow and the latter a white solid.
(n) Most of the oxysalts of lead(II) are insoluble in water, but exceptions are the nitrate and the ethanoate. The only stable oxysalt of lead(IV) is the ethanoate, but this reacts with water depositing lead(IV) oxide.

6

GROUP 5B NITROGEN, PHOSPHORUS, ARSENIC, ANTIMONY AND BISMUTH

6.1 Some Physical Data of Group 5B Elements

	Atomic number	Electronic configuration	Density/ g cm^{-3}	M.p./ K	B.p./ K	Atomic radius/nm	Ionic radius/nm M^{3+}
N	7	2.5 $1s^2 2s^2 2p^3$		25	77	0·074	
P	15	2.8.5 $...2s^2 2p^6 3s^2 3p^3$	1·83 (white P)	317	553	0·110	
As	33	2.8.18.5 $...3s^2 3p^6 3d^{10} 4s^2 4p^3$	5·73 (Grey)		886 (sublimes)	0·121	0·069
Sb	51	2.8.18.18.5 $...4s^2 4p^6 4d^{10} 5s^2 5p^3$	6·70 (Grey)	903	1910	0·141	0·090
Bi	83	2.8.18.32.18.5 $...5s^2 5p^6 5d^{10} 6s^2 6p^3$	9·80	544	1830	0·152	0·120

6.2 Some General Remarks about Group 5B

In theory, the Group 5B elements can complete the octet in chemical combination by gaining three electrons to form the 3-valent anion, by forming three covalent bonds, or by losing five electrons; the last possibility is ruled out on energetic grounds. Only nitrogen (and possibly phosphorus to a slight extent) forms the 3-valent anion, and reactive metals are required for it to be possible; the N^{3-} ion is present in ionic nitrides, e.g. $(Li^+)_3 N^{3-}$ and $(Ca^{2+})_3 (N^{3-})_2$. The majority of compounds formed by this group of elements are covalent.

Antimony and bismuth can form the 3-valent cation X^{3+} (the inert-pair effect), the Sb^{3+} ion being present in $(Sb^{3+})_2 (SO_4^{2-})_3$ and the Bi^{3+} ion in $Bi^{3+}(F^-)_3$ and $Bi^{3+}(NO_3^-)_3 \cdot 5H_2O$.

Because phosphorus, arsenic, antimony and bismuth have vacant d orbitals they are able to form 5-covalent compounds which are not possible for nitrogen, e.g. in the formation of PCl_5 one of the $3s$ electrons of the phosphorus atom is promoted to the 3d level giving five unpaired electrons for valency purposes.

Nitrogen and phosphorus are non-metallic, although one allotrope of phosphorus (black phosphorus) shows metallic conduction. Arsenic is semi-metallic and these properties become progressively more important for antimony and bismuth. Of these elements only nitrogen is able to multiple bond with itself, the triple bond being present in the nitrogen molecule, N≡N; the high stability of this molecule is the prime reason why the oxides of nitrogen are endothermic compounds. Phosphorus, arsenic and antimony are allotropic, the less dense allotrope containing X_4 tetrahedra with X—X—X bond angles of 60° which introduce considerable strain. The denser allotropic forms of phosphorus, arsenic and antimony are more stable and more 'metallic' (closer packing of the atoms). Bismuth also adopts a metallic structure.

There is little resemblance between the chemistry of nitrogen and phosphorus and these two elements are considered separately. In order to show similarities and differences, the elements arsenic, antimony and bismuth are examined together.

6.3 Extraction of the Group 5B Elements

Nitrogen occurs in the atmosphere to the extent of about 78% by volume and is extracted from the air by first liquefying it and then fractionally distilling the liquid air.

Phosphorus is extracted from calcium phosphate, $(Ca^{2+})_3(PO_4^{3-})_2$, by heating it with silica and coke in an electric furnace.

Arsenic, antimony and bismuth occur as sulphides of the general formula M_2S_3. They can be extracted by roasting in air to form the oxides which are subsequently reduced with carbon:

$$2M_2S_3 + 9O_2 \rightarrow 2M_2O_3 + 6SO_2$$

$$M_2O_3 + 3C \rightarrow 2M + 3CO$$

NITROGEN

6.4 Some Reactions of Nitrogen

Although nitrogen is rather inert it will combine with oxygen when sparked to give a mixture of nitrogen oxide and nitrogen dioxide. It also combines directly with lithium and the Group 2A metals (see Section 3.7, p. 34).

(a) Reaction of nitrogen and oxygen (Demonstration)

Using a high voltage source, pass an electric spark for a few minutes between the tips of two iron rods a centimetre apart in a loosely corked test-tube full of dry air. A brown colour should develop owing to the following two reactions:

$$N_2 + O_2 \rightarrow 2NO$$

$$2NO + O_2 \rightarrow 2NO_2 \quad \text{(nitrogen dioxide, a brown gas)}$$

(b) Reaction of magnesium and calcium with nitrogen

Carry out the experiments described in Sections 3.7(a) and 3.7(b), p. 34, if you have not previously done so.

6.5 The Hydrides of Nitrogen

Nitrogen forms three hydrides and these are ammonia, NH_3, hydrazine, N_2H_4, and hydrogen azide, HN_3.

(a) Formation of ammonia, NH_3

(i) *By the action of water on magnesium nitride and calcium nitride*

Carry out the experiments described in Section 3.7(b), p. 35, if you have not previously done so.

(ii) *By the action of sodium hydroxide solution on an ammonium salt*

To about 0.3 g of ammonium chloride in a test-tube add about 2 cm³ of approximately 2M sodium hydroxide solution and warm gently. Test for the evolution of ammonia by holding a piece of moist red litmus paper near, but not touching, the mouth of the test tube.

(b) Some properties of ammonium salts

(i) *Action of water on ammonium chloride, $NH_4^+Cl^-$*

To about 2 cm³ of distilled water (previously boiled to expel any dissolved carbon dioxide) in a test-tube add about 0.3 g of ammonium chloride. Now add 2 or 3 drops of Universal Indicator solution and compare the result with that obtained by adding the same number of drops of Universal Indicator solution to 2 cm³ of boiled-out distilled water. Can you account for the result?

See if ammonium sulphate behaves in the same way as ammonium chloride.

(ii) *Action of heat on ammonium chloride*

Place about 0.5 g of ammonium chloride in a test-tube and then further up the tube place a small plug of 'Rocksil' wool. Heat the ammonium chloride, holding a piece of moist red litmus paper near the mouth of the tube. Notice what happens to it and any further subsequent changes. Attempt to explain your observations and in particular the order in which the changes take place.

(c) Ammonia as a reducing agent—oxidation with copper(II) oxide

Place about 0·3 g of ammonium chloride in a test-tube and then mix it thoroughly with about five times its own volume of soda-lime. Place a small plug of 'Rocksil' wool above this mixture and above this place a small pile of copper(II) oxide. Heat the test-tube near the wool plug so that the ammonium chloride/soda-lime and copper(II) oxide are both heated and observe any changes that occur to the copper(II) oxide. Attempt to write an equation for the reaction.

(d) Action of water on hydrazine, NH_2—NH_2

Use hydrazine hydrate, $N_2H_4 \cdot H_2O$, for this experiment. Add about 3 drops of hydrazine hydrate to about 2 cm^3 of distilled water in a test-tube and then add 2 or 3 drops of Universal Indicator solution. What happens and why?

(e) Hydrazine as a reducing agent

Hydrazine is an endothermic compound which burns in air to give nitrogen and steam with a vigorous evolution of heat. This fact accounts for the interest shown in it as a potential rocket fuel. In the following experiments hydrazine hydrate (see above) is used.

(i) *Reduction of iodine solution*

Dissolve about 3 drops of hydrazine hydrate in 2 cm^3 of distilled water and then add, drop by drop, a dilute solution of iodine in potassium iodide solution (the purpose of the potassium iodide is simply to get sufficient iodine to dissolve). Record what happens. Do you notice any bubbles of gas being evolved? If so, what do you think this gas is?

Since reduction is gain of electrons and oxidation is loss of electrons, attempt to write down two partial equations to show what is happening in this reaction, e.g.

$$NH_2-NH_2 \rightarrow ? + ? + 4e^-$$
$$2I_2 + 4e^- \rightarrow$$

(ii) *Reduction of silver nitrate solution*

Make a dilute solution of hydrazine hydrate as in (i) above and then add a few drops of dilute silver nitrate solution. Observe what happens and then attempt to write down two partial equations for the reaction, as in (i) above.

(iii) *Reduction of acidified potassium permanganate solution*

Make a dilute solution of hydrazine hydrate as in (ii) above, and then add a few drops of a dilute solution of potassium permanganate, previously acidified with a little dilute sulphuric acid. Observe what happens. Given the following information:

$$MnO_4^- + 8H^+ + 5e^- \rightarrow Mn^{2+} + 4H_2O$$

$$NH_2-NH_2 \rightarrow ? + ? + 4e^-$$

complete the above two partial equations and then write a balanced chemical equation for the overall reaction, eliminating the electrons from the two partial equations.

(f) **Action of water on hydrazine sulphate, $(NH_3^+-NH_3^+)(SO_4^{2-})$**

To about $2 cm^3$ of distilled water (previously boiled to expel any dissolved carbon dioxide) in a test-tube add about $0.3 g$ of hydrazine sulphate. Now add 2 or 3 drops of Universal Indicator solution and compare the result with that obtained by adding the same number of drops of Universal Indicator solution to $2 cm^3$ of boiled-out distilled water. Can you account for the result?

6.6 The Oxides of Nitrogen

The best known oxides of nitrogen are dinitrogen oxide, N_2O, nitrogen oxide, NO, dinitrogen trioxide, N_2O_3, nitrogen dioxide, NO_2 in equilibrium with N_2O_4, and dinitrogen pentoxide, N_2O_5. They are all endothermic compounds and this is primarily due to the high stability of the nitrogen molecule.

Nitrogen dioxide can be obtained by the action of concentrated nitric acid on copper or by heating most ionic nitrates. Copper and 50% nitric acid produces mainly nitrogen oxide, while the action of dilute nitric acid on zinc produces some dinitrogen oxide. Dinitrogen trioxide only exists in the pure state at low temperatures and is about 90% dissociated into nitrogen dioxide and nitrogen oxide at room temperature. Dinitrogen pentoxide is a colourless solid obtained by dehydrating nitric acid with phosphorus(V) oxide.

6.7 The Oxyacids of Nitrogen and their Salts

The best known oxyacids of nitrogen are nitrous acid, HNO_2, and nitric acid, HNO_3. They give rise to nitrites and nitrates respectively.

(a) Formation and unstable nature of an aqueous solution of nitrous acid, HNO_2

To about 0·3 g of sodium nitrite in a test-tube add about 2 cm³ of ice cold approximately M sulphuric acid. Observe what happens and then allow the mixture to warm to room temperature.

(b) Oxidising action of nitrous acid and acidified solutions of nitrites

(i) *Oxidation of iron(II) sulphate solution*

Dissolve about 0·1 g of sodium nitrite in 2 cm³ of distilled water and then add about 2 cm³ of approximately M sulphuric acid. To this mixture (which contains nitrous acid) add a few drops of a dilute solution of iron(II) sulphate solution. Heat the mixture until the colour lightens and then allow it to cool. Then add approximately 2M sodium hydroxide solution gradually and note the colour of the precipitate formed. What is it? Add sodium hydroxide solution to some of the iron(II) sulphate solution and note the colour of the precipitate. What has happened to the iron(II) ions in this experiment?

(ii) *Oxidation of potassium iodide solution*

Prepare an aqueous solution of nitrous acid as in (i) above, and then add a few drops of a dilute solution of potassium iodide. What happens to the iodide ions?

When nitrous acid (and acidified solutions of nitrites) act as oxidising agents, the nitrite ion is reduced as below:

$$4H^+ + 2NO_2^- + 2e^- \rightarrow 2H_2O + 2NO$$

Using the above information, write balanced equations for the action of acidified solutions containing the nitrite ion with iron(II) ions and with iodide ions.

(c) Reducing action of nitrous acid and acidified solutions of nitrites

Reduction of acidified potassium permanganate solution

Dissolve about 0·1 g of sodium nitrite in 2 cm³ of distilled water and then add about 2 cm³ of approximately M sulphuric acid. To this mixture add a few drops of a very dilute solution of potassium permanganate and observe what happens.

Using the information below, construct balanced partial equations showing loss and gain of electrons and then construct a balanced equation with the electrons eliminated.

$$MnO_4^- + H^+ + electrons \rightarrow Mn^{2+} + 4H_2O$$

$$H_2O + NO_2^- \rightarrow NO_3^- + H^+ + electrons$$

(d) Tests for nitrites

Experiments 6.7(a) and 6.7(b)(i) can be used as tests for the nitrite ion, but in 6.7(b)(i) the solution should not be acidified.

(e) Action of dilute nitric acid, HNO_3, on a selection of metals, oxides and carbonates

Examine the effect of small amounts of approximately 2M nitric acid on small quantities of copper, copper(II) oxide and copper(II) carbonate. Warm gently if this is necessary. Repeat with other substances, e.g. zinc and lead, and their oxides and carbonates.

It is said that very dilute nitric acid reacts with magnesium to give hydrogen. See if you can verify this.

(f) Oxidising action of nitric acid, HNO_3

(i) *Oxidation of iron(II) sulphate solution*

To about 2 cm^3 of a dilute solution of iron(II) sulphate add 2 or 3 drops of concentrated nitric acid and warm for a minute or two. Allow the solution to cool, and then add approximately 2M sodium hydroxide solution gradually, until a permanent precipitate is obtained. Does nitric acid behave in a similar way to nitrous acid in this experiment?

(ii) *Oxidation of potassium iodide solution*

To about 2 cm^3 of a dilute solution of potassium iodide add 2 or 3 drops of concentrated nitric acid. Observe and explain the result.

(g) Passivity of iron

Concentrated nitric acid has no effect on aluminium, iron or chromium. It is thought that the surface of these three metals is first attacked, a thin impervious oxide film being formed which stops further action; the phenomenon is referred to as passivity.

Clean a small iron nail with sandpaper and suspend it for a few seconds in a small beaker containing concentrated nitric acid. Wash it in distilled water and then transfer it to a solution of copper(II) sulphate. There should be no coating of copper on the surface of the nail (if there is, try a different nail).

While the copper(II) sulphate is still adhering to the nail, pinch it with the fingers or drop it on to the bench; a layer of copper should immediately form, showing that the protective film has been ruptured. Place an untreated nail in copper(II) sulphate solution; displacement of copper should immediately be apparent.

(h) Action of heat on nitrates

Heat about 0.3 g of potassium nitrate strongly in an ignition tube and test for oxygen with a glowing splint. Continue heating strongly for a few minutes and then allow the residue to cool. Now add about $2\,cm^3$ of approximately M sulphuric acid to the residue and note the result. Try the effect of approximately M sulphuric acid on an untreated sample of potassium nitrate and notice if there is any difference. In view of what you have observed, what do you think is the solid residue when potassium nitrate is heated?

Repeat the experiment with similar small portions of barium nitrate, copper(II) nitrate and lead(II) nitrate but in these cases do not add the dilute sulphuric acid. How does the action of heat on these nitrates differ from the action of heat on potassium nitrate? Can you suggest a reason?

(i) Test for nitrates

To about $0.3\,g$ of potassium nitrate in a test-tube add about $2\,cm^3$ of approximately M sulphuric acid. Is there any apparent reaction? Does this experiment distinguish a nitrite from a nitrate?

Dissolve about $0.3\,g$ of potassium nitrate in about $3\,cm^3$ of distilled water and add the same volume of a freshly prepared solution of iron(II) sulphate solution. Cool the mixture under running water and **cautiously** add some concentrated sulphuric acid to form a lower layer. Notice the formation of a brown ring at the junction of the two liquids which is due to the production of the addition compound $FeSO_4.NO$.

Show that other nitrates give the same test, for example, aqueous solutions of magnesium nitrate and zinc nitrate.

PHOSPHORUS

6.8 Allotropy and Some Reactions of Phosphorus

(a) Conversion of white phosphorus into red phosphorus (Demonstration)

Half fill a test-tube with water and then quickly add a small quantity of white phosphorus and a small crystal of iodine. Heat the tube gently for about half an hour but do not allow the water to boil. The white phosphorus should now have changed into the more stable red variety. Iodine catalyses the conversion.

(b) Conversion of red phosphorus into white phosphorus (Demonstration)

Assemble the apparatus shown in Fig. 6.1, preferably in a fume cupboard. Pass carbon dioxide through the boiling-tube until all the air has been expelled and then heat very gently to about 45°C. White phosphorus should be seen to condense on the cooler surfaces of the boiling-tube.

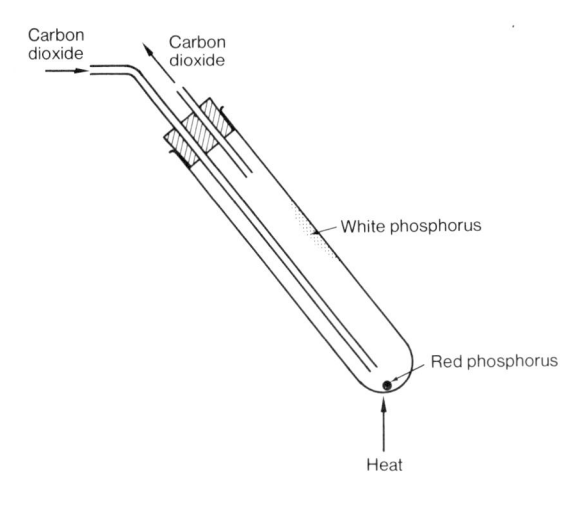

Fig. 6.1
Conversion of red phosphorus into white phosphorus.

(c) **Difference in reactivity between white and red phosphorus (Demonstration)**

Carry out this experiment in a fume cupboard. On to one end of a strip of iron about 15 cm in length place a small pile of red phosphorus and at the other end place a small piece of white phosphorus. Heat the iron strip somewhat off-centre and nearer to the red phosphorus. Which variety of phosphorus inflames more readily?

(d) **Action of carbon disulphide on red and white phosphorus (Demonstration)**

Caution: carbon disulphide is very poisonous and highly flammable so carry out this experiment in a fume cupboard. Place about $3\,cm^3$ of carbon disulphide into each of two separate test-tubes and then add a small quantity of red phosphorus to one tube and a small piece of white phosphorus to the other. Which allotrope of phosphorus dissolves?

Pour the solution of phosphorus in carbon disulphide on to a filter paper. **Caution: do not hold the filter-paper in the hand.** Allow the carbon disulphide to evaporate of its own accord and note the result.

Attempt to explain the results of Experiments 6.8(c) and (d) in terms of the structure of red and white phosphorus. Consult an inorganic textbook to check your answer.

6.9 The Hydrides of Phosphorus

Phosphorus forms two hydrides, namely phosphine, PH_3, and diphosphane, P_2H_4, the phosphorus analogues of ammonia and hydrazine respectively.

Formation of phosphine, PH_3 (Demonstration)

Carry out this experiment in a fume cupboard. Place a small lump of calcium phosphide in a small evaporating dish and **cautiously** squirt into it a small amount of water. Notice that a vigorous reaction takes place with the liberation of a gas (phosphine), which spontaneously bursts into flames. The flammable nature of the gas is said to be due to the presence of diphosphane, P_2H_4, impurities.

Compare this reaction with the action of water on calcium nitride or magnesium nitride carried out previously:

$$P^{3-} + 3H_2O \rightarrow PH_3 + 3OH^-$$

$$N^{3-} + 3H_2O \rightarrow NH_3 + 3OH^-$$

Phosphine, unlike ammonia, is only slightly soluble in water. It is also less basic than ammonia but it does form phosphonium iodide, $PH_4^+I^-$; however this salt decomposes more readily than ammonium salts:

$$PH_4^+I^- \rightarrow PH_3 + HI$$

6.10 The Oxides of Phosphorus

There are two main oxides of phosphorus, namely phosphorus(III) oxide, once thought to be P_2O_3 but now known to be a dimer (m.p. 297K), and phosphorus(V) oxide a white solid which is also dimeric. They can both be obtained by burning phosphorus in oxygen under the right conditions. Their structures are based on the tetrahedral P_4 molecule and are given below:

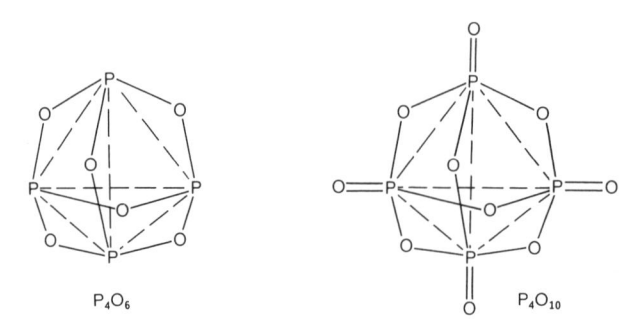

P_4O_6 P_4O_{10}

Fig. 6.2   Fig. 6.3
The structure of phosphorus(III) oxide.   The structure of phosphorus(V) oxide.

Notice that the phosphorus atom is 3-covalent in P_4O_6 but 5-covalent in P_4O_{10}.

The action of water on phosphorus(V) oxide, P_4O_{10}

To about $2\,cm^3$ of distilled water in a test-tube add cautiously very small portions of phosphorus(V) oxide a little at a time and note the vigour of the reaction. Test the resulting solution with a few drops of Universal Indicator solution. Now boil the solution for a short while and save the solution so that you can carry out the test for the orthophosphate anion (see later).

Phosphorus(V) oxide reacts with water forming metaphosphoric acid (empirical formula HPO_3) which on boiling gives orthophosphoric acid, H_3PO_4:

$$P_4O_{10} + 2H_2O \rightarrow 4HPO_3$$
$$HPO_3 + H_2O \rightarrow H_3PO_4$$

6.11 The Oxyacids of Phosphorus and their Salts

The main oxyacids of phosphorus are the phosphorus acids and the phosphoric acids, of which phosphorus(III) oxide and phosphorus(V) oxide respectively are the acid anhydrides. Other oxyacids of phosphorus exist, but they are not based directly on these oxides and will not be considered.

(a) Phosphorous acids and phosphites

Phosphorus forms many phosphorous acids and, except for orthophosphorous acid, H_3PO_3, they are polymeric. Their structures contain —P—O—P— bonds, c.f. the silicic acids based on —Si—O—Si— bonds (p. 58). Only orthophosphorous acid and disodium hydrogen orthophosphite, $(Na^+)_2HPO_3^{2-}$, are considered.

(i) *Orthophosphorous acid, H_3PO_3, as a reducing agent*

Dissolve about $0.3\,g$ of orthophosphorous acid in about $6\,cm^3$ of distilled water and divide the solution into three portions. To one portion add, drop by drop, a dilute solution of iodine in potassium iodide solution. To the second portion add a few drops of dilute silver nitrate solution. To the third portion add a few drops of a dilute solution of potassium permanganate, previously acidified with dilute sulphuric acid. Comment on your results.

Attempt to draw the molecular structure of the orthophosphorous acid molecule which contains 5-valent phosphorus. Check your answer by consulting an inorganic textbook.

(ii) *The action of water on disodium hydrogen orthophosphite,* $(Na^+)_2HPO_3^{2-}$

Dissolve about 0·1 g of the solid in about 3 cm³ of distilled water and then add 2 or 3 drops of Universal Indicator solution. Repeat the experiment with a solution of sodium chloride in place of the orthophosphite and compare the results.

Attempt to explain your results by considering the effect of water on the HPO_3^{2-} ion. Do you consider orthophosphorous acid to be stronger or weaker than hydrochloric acid? Give reasons.

(b) Phosphoric acids and phosphates

Like the phosphorous acids, a large variety of phosphoric acids exists whose structures contain —P—O—P— bonds. In many respects the structural chemistry of these acids and their anions is similar to that of the silicates (p. 58).

(i) *Formation of orthophosphoric acid,* H_3PO_4

This may be obtained by the action of water on phosphorus(V) oxide. Orthophosphoric acid is obtained on boiling the solution (see Section 6.10, p. 83).

The pure acid is a deliquescent solid but it is usually encountered as a rather viscous liquid. Attempt to draw the molecular structure of the orthophosphoric acid molecule which contains 5-valent phosphorus.

(ii) *Orthophosphoric acid is devoid of oxidising properties*

Dissolve a little syrupy orthophosphoric acid in about 2 cm³ of distilled water and add a few drops of a dilute solution of potassium iodide. Is there any noticeable reaction?

(iii) *Action of water on sodium dihydrogen orthophosphate,* $Na^+H_2PO_4^-$, *and disodium hydrogen orthophosphate,* $(Na^+)_2HPO_4^{2-}$

Dissolve about 0·2 g of sodium dihydrogen orthophosphate in about 5 cm³ of distilled water and divide into two portions. To one portion add 3 drops of phenolphthalein (colour change from colourless to red occurs over the pH range 8·3–10) and to the other portion add 3 drops of bromothymol blue indicator solution (colour change from yellow to blue occurs over the pH range 6–7·6). Record the indicator colours. Repeat the experiment, but this time use a solution of disodium hydrogen orthophosphate. Attempt to explain your results by considering the effect of water on the $H_2PO_4^-$ and HPO_4^{2-} ions.

(iv) *Test for the orthophosphate ion*

Dissolve about 0·1 g of sodium dihydrogen orthophosphate in about 1 cm³ of distilled water and add about 5 cm³ of ammonium molybdate solution followed by about 5 drops of concentrated nitric acid. Warm the solution and note the formation of a yellow precipitate of ammonium phosphomolybdate. This is a test for the orthophosphate ion.

Repeat with aqueous solutions of disodium hydrogen orthophosphate, a dilute solution of orthophosphoric acid and the boiled solution from Experiment 6.10, p. 83.

(v) *Formation of tetrasodium pyrophosphate,* $(Na^+)_4P_2O_7^{4-}$, *a salt of pyrophosphoric acid,* $H_4P_2O_7$

Heat about 0·3 g of disodium hydrogen orthophosphate in a test-tube until no further evolution of steam occurs and then allow the residue to cool. The solid is tetrasodium pyrophosphate, obtained by the elimination of water thus:

$$2(Na^+)_2HPO_4^{2-} \rightarrow (Na^+)_4P_2O_7^{4-} + H_2O$$

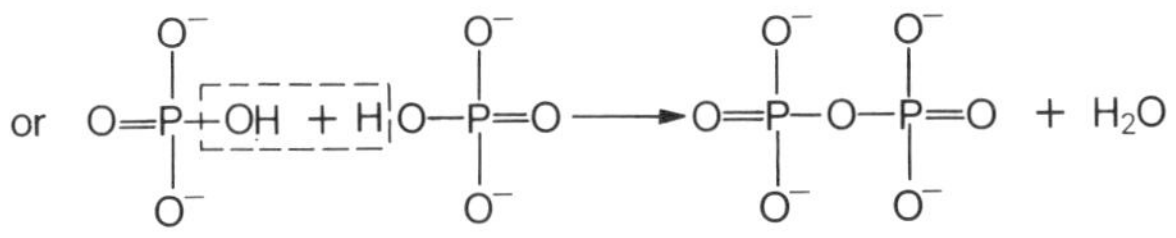

Pyrophosphoric acid and pyrophosphates give the yellow precipitate of ammonium phosphomolybdate, but at a slower rate than orthophosphates, since hydrolysis occurs on warming with water to give the orthophosphate ion:

$$P_2O_7^{4-} + H_2O \rightarrow 2HPO_4^{2-}$$

Carry out this test on the solid you have just made according to the method given in Section 6.11(b)(iv). You should observe that the solution needs rather more heating than before.

(vi) *Formation of metaphosphoric acid,* HPO_3

Metaphosphoric acid is a glassy polymeric solid and a mixture of acids with the empirical formula HPO_3. Although the structures of some of the metaphosphoric acids are still uncertain, they are known to be cyclic as opposed to linear and are built up from PO_4 units. The trimetaphosphate ion, $P_3O_9^{3-}$, is shown below:

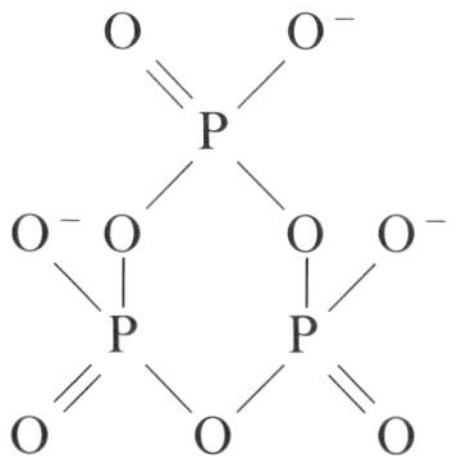

Heat about 0·5 cm³ of syrupy orthophosphoric acid in a test-tube until evolution of steam ceases and then allow the residue to cool. Examine the product by extracting a little on a glass rod. Can you explain why the product is glassy rather than a crystalline solid?

$$H_3PO_4 \rightarrow HPO_3 + H_2O$$

Dissolve the residue in about 2 cm³ of distilled water and treat it with ammonium molybdate solution and a few drops of concentrated nitric acid as in Experiment 6.11(b)(iv). Does precipitation of yellow ammonium phosphomolybdate occur as readily as with an orthophosphate? Explain your results in terms of an equation showing the action of water on HPO_3.

(vii) *Formation of sodium metaphosphate*, $Na^+PO_3^-$

Sodium metaphosphate, empirical formula $Na^+PO_3^-$ is a mixture containing cyclic metaphosphate ions of differing ring size.

Heat about 0·3 g of sodium dihydrogen orthophosphate in a test-tube until evolution of steam ceases and no further change occurs. Allow the residue to cool and note whether it appears crystalline or glass-like. Can you explain its appearance?

$$Na^+H_2PO_4^- \rightarrow Na^+PO_3^- + H_2O$$

Dissolve the residue in about 2 cm³ of distilled water and treat it with ammonium molybdate solution and a few drops of concentrated nitric acid as in Experiment 6.11(b)(iv). How readily does the yellow precipitate of ammonium phosphomolybdate appear on warming? Explain your results.

6.12 The Chlorides of Phosphorus

Phosphorus forms two chlorides, phosphorus trichloride, PCl_3, and phosphorus pentachloride, PCl_5. The former is a colourless liquid and the latter a white solid.

(a) Action of water on phosphorus trichloride, PCl_3

Place about 5 drops of phosphorus trichloride in a test-tube and **cautiously** add a little distilled water. Does the tube become warm? Heat gently to complete the reaction. Add 2 or 3 drops of Universal Indicator solution and comment on the result. Attempt to write an equation for the reaction of phosphorus trichloride with water.

(b) Some reactions of phosphorus pentachloride, PCl_5

(i) *Action of heat on phosphorus pentachloride*

Heat about 0·2 g of phosphorus pentachloride in a test-tube and hold a piece of filter paper strip soaked in potassium iodide solution near the mouth of the tube. Observe and explain what happens. Attempt to write an equation for the action of phosphorus pentachloride on iodide ions.

(ii) *Action of water on phosphorus pentachloride*

Place about 0·1 g of phosphorus pentachloride in a test-tube and **cautiously** add about 2 cm³ of distilled water. Does the tube become warm? Observe what happens. Test the resulting solution for acidity with Universal Indicator solution and show that the orthophosphate ion is formed by carrying out the test given in Section 6.11(b)(iv). Attempt to write an equation for the action of water on phosphorus pentachloride.

You should have noticed that hydrogen chloride was evolved (a gas which fumes in moist air). Organic compounds which contain the —OH group react with phosphorus pentachloride evolving hydrogen chloride; this is a specific test for the presence of an —OH group in such compounds.

(iii) *Action of ethanol on phosphorus pentachloride*

Place about 0·1 g of phosphorus pentachloride in a test-tube and **cautiously** add a little ethanol drop by drop. You should notice the evolution of a fuming gas (hydrogen chloride):

$$C_2H_5OH + PCl_5 \rightarrow C_2H_5Cl + POCl_3 + HCl$$

ARSENIC, ANTIMONY AND BISMUTH

6.13 Structures and some Reactions of Arsenic, Antimony and Bismuth

Arsenic and antimony exhibit allotropy, an unstable form of each element structurally similar to white phosphorus being formed by the rapid condensation of their vapours. This unstable form is readily transformed into a denser allotrope which is metallic and similar in structure to that of black phosphorus, i.e. it is macromolecular. Bismuth does not exhibit allotropy and adopts the latter structure.

The elements combine, on heating, with oxygen, sulphur and the halogens. Concentrated sulphuric acid attacks antimony and bismuth forming the sulphates (indicative of metallic behaviour), while arsenic is converted into the oxide As_4O_6 (an indication of non-metallic behaviour).

Arsenic and its compounds in particular are exceedingly poisonous and extreme caution should be exercised when handling them.

6.14 The Hydrides of Arsenic, Antimony and Bismuth, MH_3

Arsenic, antimony and bismuth form gaseous hydrides of the general formula MH_3 which are increasingly unstable in this order, i.e. they become more powerful reducing agents. It was noted earlier that phosphine was less basic than ammonia; this trend continues and none of these hydrides form ions analogous to NH_4^+ or PH_4^+.

Bismuthine, BiH_3, is extremely unstable with respect to its elements and was first detected by treating a bismuth-magnesium alloy, containing radioactive bismuth, with dilute acid. The detection of radioactivity in the gas phase showed that a volatile hydride of bismuth existed.

6.15 The Oxides of Arsenic, Antimony and Bismuth

(a) The 3-valent oxides

Arsenic and antimony(III) oxides are obtained when the respective metals are heated in oxygen. Arsenic(III) oxide exists in two polymorphic forms one of which is structurally similar to phosphorus(III) oxide (p. 82) having the formula As_4O_6. Antimony(III) oxide, Sb_4O_6, has a similar structure. Bismuth(III) oxide, obtained by heating the carbonate or nitrate, exists in several forms, one of which is ionic and thus represented by the empirical formula $(Bi^{3+})_2(O^{2-})_3$.

(i) *Action of water on the 3-valent oxides of arsenic, antimony and bismuth*

Take about 0·1 g of arsenic(III) oxide in a test-tube and add about 2 cm³ of distilled water. Shake and then add 2 or 3 drops of Universal Indicator solution. Note and explain the result.

Repeat the experiment with antimony(III) oxide and bismuth(III) oxide and comment on your results.

(ii) *Action of sodium hydroxide solution on the 3-valent oxides of arsenic, antimony and bismuth*

Take about 0·1 g of each oxide in separate test-tubes and add to each about 2 cm³ of approximately 2M sodium hydroxide solution. Warm gently and note which oxides dissolve. From the results of this and the previous experiment can you classify these three oxides?

(b) The 5-valent oxides

Arsenic and antimony(V) oxides may be obtained by oxidation of the elements with concentrated nitric acid. Since their structures are unknown (unlike the structure of phosphorus(V) oxide) they are represented by the respective empirical formulae As_2O_5 and Sb_2O_5.

It is possible that bismuth(V) oxide exists, although it has never been obtained in the pure form.

(i) *Action of water on the 5-valent oxides of arsenic and antimony*

Take about 0·1 g of each oxide in separate test-tubes and add to each about 2 cm³ of distilled water. Shake the tubes and then add 2 or 3 drops of Universal Indicator solution to each. Note and explain your observations.

(ii) *Action of sodium hydroxide solution on the 5-valent oxides of arsenic and antimony*

Take about 0·1 g of each oxide in separate test-tubes and add to each about 2 cm³ of approximately 2M sodium hydroxide solution. Warm gently and comment on your observations. What type of oxides are these?

6.16 The Oxyacids of Arsenic and Antimony and their Salts

There is some doubt about the nature of arsenious acid and it is thought that the action of water on arsenic(III) oxide gives the hydrated oxide. However, the orthoarsenite, AsO_3^{3-}, and the meta-arsenite ion (empirical formula AsO_2^-) are formed on treating arsenic(III) oxide with hot alkali. Antimonous acid does not exist but antimonites containing the meta-antimonite ion (empirical formula SbO_2^-) are known.

(a) **Reducing action of sodium arsenite, $(Na^+)_3AsO_3^{3-}$**

(i) *Action of sodium arsenite on iodine solution*

To about 2 cm³ of a solution of sodium arsenite add a few drops of a dilute solution of iodine in potassium iodide solution. Observe the result.

(ii) *Action of sodium arsenite on an acidified solution of potassium permanganate*

To about 2 cm³ of a solution of sodium arsenite add a few drops of a dilute solution of potassium permanganate previously acidified with dilute sulphuric acid and note the result.

Arsenic acid can be crystallised from water as a white hydrated solid, $H_3AsO_4 \cdot \frac{1}{2}H_2O$. It is a tribasic acid, forming salts containing the ortho-arsenate ion, AsO_4^{3-}, which are generally isomorphous with the corresponding orthophosphates. Antimonic acid has never been isolated in the pure state, although solid antimonates such as $K^+[Sb(OH)_6]^-$ are known (note that the co-ordination number of antimony in antimonates is six, whereas phosphorus and arsenic have co-ordination numbers of four in phosphates and arsenates). There is no higher acid of bismuth corresponding to arsenic acid and although sodium bismuthate, $Na^+BiO_3^-$, exists it has never been obtained pure.

(b) **Some reactions of sodium arsenate, $(Na^+)_3AsO_4^{3-}$**

(i) *Oxidising action of sodium arsenate*

To about 2 cm³ of a solution of sodium arsenate add an equal volume of concentrated hydrochloric acid and then add a solution of potassium iodide drop by drop. Note the liberation of iodine.

In view of the reduction of an iodine solution with sodium arsenite this result appears surprising. However, the reaction is reversible and is as indicated below:

$$AsO_4^{3-} + 2I^- + H_2O \rightleftharpoons AsO_3^{3-} + I_2 + 2OH^-$$

Can you see why the solution was acidified with concentrated hydrochloric acid in the above oxidation?

(ii) *Test for the orthoarsenate ion*

To about 1 cm³ of a solution of sodium arsenate add about 5 cm³ of ammonium molybdate solution, followed by about 5 drops of concentrated nitric acid. Heat the solution and note the formation of a yellow precipitate of ammonium arsenomolybdate.

Does this precipitation occur more or less readily than the similar reaction with an orthophosphate (p. 84)?

(c) **Oxidising action of sodium bismuthate, $Na^+BiO_3^-$**

Add about 0·2 g of sodium bismuthate to a cold solution of manganese(II) sulphate containing some approximately 2M nitric acid. Stir with a glass rod and then filter the mixture. A purple colour of permanganic acid should be seen, i.e. the sodium bismuthate has oxidised manganese(II) to manganese(VII).

6.17 Bismuth Hydroxide, $Bi(OH)_3$

The hydroxide of bismuth is exclusively basic and this is indicative of the metallic properties of this element.

Formation of bismuth hydroxide

Place about 0·1 g of bismuth nitrate in a test-tube and then add about 2 cm³ of distilled water. At this stage the mixture will be milky, so add concentrated nitric acid drop by drop until a clear solution is obtained. Now add approximately 2M sodium hydroxide solution uhtil a precipitate forms:

$$Bi^{3+} + 3OH^- \rightarrow Bi(OH)_3$$

6.18 The Sulphides of Arsenic, Antimony and Bismuth

Arsenic forms four sulphides but the most common ones are arsenic(III) sulphide and arsenic(V) sulphide with formulae respectively As_2S_3 and As_2S_5. Antimony forms two sulphides, antimony(III) sulphide, Sb_2S_3, and antimony(V) sulphide, Sb_2S_5. Bismuth forms only the 3-valent sulphide Bi_2S_3.

(a) Formation of arsenic(III) sulphide, As$_2$S$_3$

Place about 2 cm^3 of a solution of sodium arsenite (contains arsenic(III)) in a test-tube and then add about the same volume of concentrated hydrochloric acid. Pass a stream of hydrogen sulphide through the solution, **in a fume cupboard,** and note the colour of the precipitate:

$$2AsO_3^{3-} + 6H^+ + 3H_2S \rightarrow As_2S_3 + 6H_2O$$

Add approximately 2M sodium hydroxide solution to the mixture and note whether the sulphide precipitate dissolves.

(b) Formation of arsenic(V) sulphide, As$_2$S$_5$

Place about 2 cm^3 of a solution of sodium arsenate (contains arsenic(V)) in a test-tube and then add about the same volume of concentrated hydrochloric acid. Warm the solution and then pass a stream of hydrogen sulphide through it. Allow to stand, and, if no precipitate forms, continue to pass hydrogen sulphide into the mixture. Note the colour of the precipitate:

$$2AsO_4^{3-} + 6H^+ + 5H_2S \rightarrow As_2S_5 + 8H_2O$$

Add approximately 2M sodium hydroxide solution to the mixture and note whether the sulphide precipitate dissolves.

(c) Formation of antimony(III) sulphide, Sb$_2$S$_3$

Place about 0·1 g of antimony trichloride in a test-tube and add about 2 cm^3 of approximately 2M hydrochloric acid and then sufficient concentrated hydrochloric acid, drop by drop, until a clear solution is obtained. Pass a stream of hydrogen sulphide through the solution and note the colour of the precipitate:

$$2SbCl_3 + 3H_2S \rightarrow Sb_2S_3 + 6HCl$$

Add approximately 2M sodium hydroxide solution to the mixture and note whether the sulphide precipitate dissolves.

(d) Formation of antimony(V) sulphide, Sb$_2$S$_5$

Place about 0·1 g of potassium antimonate in a test-tube (contains antimony(V)) and dissolve it in about 2 cm^3 of distilled water. Pass a stream of hydrogen sulphide through the solution and note the colour of the precipitate:

$$2Sb(OH)_6^- + 5H_2S \rightarrow Sb_2S_5 + 10H_2O + 2OH^-$$

Add approximately 2M sodium hydroxide solution to the mixture and note whether the sulphide precipitate dissolves.

(e) Formation of bismuth(III) sulphide, Bi$_2$S$_3$

Place about 0·1 g of bismuth nitrate in a test-tube and then add about 2 cm^3 of distilled water. Now add sufficient nitric acid, drop by drop, to obtain a clear solution. Pass a stream of hydrogen sulphide through the solution and note the colour of the precipitate:

$$2Bi^{3+} + 3H_2S \rightarrow Bi_2S_3 + 6H^+$$

Add approximately 2M sodium hydroxide solution to the mixture and note whether the sulphide precipitate dissolves.

Summarise the behaviour of the sulphides of arsenic, antimony and bismuth towards sodium hydroxide solution. In view of the fact that oxygen and sulphur are in the same Group in the Periodic Table, it might be expected that these sulphides would behave like the corresponding oxides. Which sulphides are showing 'acidic' properties and which are showing 'basic' properties? Is this what you would expect?

6.19 The Chlorides of Arsenic, Antimony and Bismuth

(a) The 3-valent chlorides

Arsenic trichloride, $AsCl_3$, is a liquid which is hydrolysed by water with the formation of hydrated arsenic(III) oxide:

$$4AsCl_3 + 6H_2O \rightarrow As_4O_6 + 12HCl$$

Antimony trichloride, $SbCl_3$, and bismuth trichloride, $BiCl_3$, are solids and incompletely hydrolysed by water with the formation of oxysalts (a type of behaviour indicative of the increased metallic character of antimony and bismuth):

$$SbCl_3 + H_2O \rightleftharpoons SbOCl + 2HCl$$

$$BiCl_3 + H_2O \rightleftharpoons BiOCl + 2HCl$$

Action of water on antimony trichloride, $SbCl_3$, and on bismuth trichloride, $BiCl_3$

Place about 0·1 g of antimony trichloride in a test-tube and then add about 2 cm^3 of approximately 2M hydrochloric acid, and then sufficient concentrated hydrochloric acid, drop by drop, until a clear solution is obtained. Dilute this solution with distilled water and note the formation of a precipitate of antimony oxychloride (antimony(III) chloride oxide), SbOCl. Dissolve this in the minimum amount of concentrated hydrochloric acid and then dilute again with water, to establish that this reaction is easily reversible.

Repeat the experiment exactly with bismuth oxychloride (bismuth(III) chloride oxide) (the addition of concentrated hydrochloric acid converts it into bismuth trichloride).

(b) The 5-valent chlorides

The only pentachloride known to exist is antimony pentachloride, $SbCl_5$, which is a liquid and a powerful oxidising agent, readily decomposing into antimony trichloride and chlorine:

$$SbCl_5 \rightarrow SbCl_3 + Cl_2$$

6.20 Some Salts of Antimony and Bismuth

There is no evidence that arsenic forms a cation, however the more metallic antimony and bismuth can form respectively the Sb^{3+} and Bi^{3+} cations, in which a pair of electrons remain inert (the inert-pair effect). These cations occur in $(Sb^{3+})_2(SO_4^{2-})_3$, $(Bi^{3+})_2(SO_4^{2-})_3$, $Bi^{3+}(NO_3^-)_3$ and $Bi^{3+}(F^-)_3$; bismuth because of its greater 'metallic character' forms more salts than antimony.

The action of heat on some bismuth salts

(i) *The action of heat on basic bismuth carbonate,* $(BiO^+)_2CO_3^{2-}$

Heat about 0·3 g of the solid in an ignition tube and test for carbon dioxide in the usual way.

(ii) *The action of heat on bismuth nitrate,* $Bi^{3+}(NO_3^-)_3$

Heat about 0·3 g of the solid in an ignition tube and identify the two gases evolved. Attempt to write an equation for this reaction.

6.21 Summary

(a) Nitrogen is rather unreactive but it will react with oxygen when the mixed gases are sparked to give a mixture of nitrogen oxide and nitrogen dioxide. It will also combine directly with lithium and the Group 2A metals on heating.

(b) The important hydrides of nitrogen are ammonia, NH_3, and hydrazine, N_2H_4. Ammonia may be obtained by heating an ammonium salt with an alkali. It is a reducing agent; for example, it will reduce hot copper(II) oxide to copper, and with water it produces a weakly alkaline solution (aqueous ammonia). Ammonium salts of strong acids, e.g. ammonium chloride, give an acidic reaction with water. Hydrazine is an endothermic compound and a reducing agent; for example, hydrazine hydrate reduces iodine solution to iodide ions, silver ions to silver, and permanganate ions to manganese(II) ions, being itself converted into nitrogen and hydrogen ions. Like ammonia it forms salts, and hydrazine sulphate shows an acidic reaction in water.

(c) The best known oxides of nitrogen are dinitrogen oxide, N_2O, nitrogen oxide, NO, dinitrogen trioxide, N_2O_3, nitrogen dioxide, NO_2, and dinitrogen pentoxide, N_2O_5. All are endothermic compounds.

(d) The best known oxyacids of nitrogen are nitrous acid, HNO_2, and nitric acid, HNO_3; they give rise to nitrites and nitrates respectively. Nitrous acid is unstable, readily decomposing into water, nitrogen oxide and nitrogen dioxide; nitrous acid and acidified solutions of nitrites are generally oxidising agents, but potassium permanganate forces them to assume the rôle of reducing agents. Nitric acid is a strong acid in aqueous solution and both the concentrated and dilute acids function as oxidising agents. Nitrites, but not nitrates, evolve nitrogen dioxide when treated with dilute hydrochloric acid.

(e) Phosphorus exhibits allotropy—white and red. White phosphorus is more chemically reactive than the red allotrope.

(f) Phosphine, PH_3, is the phosphorus analogue of ammonia but it is far less basic than ammonia although it will form a salt, phosphonium iodide, $PH_4^+I^-$. The hydride may be obtained by the action of water on calcium phosphide as a spontaneously flammable gas (owing to the presence of diphosphane, P_2H_4). The pure gas is not spontaneously flammable.

(g) There are two main oxides of phosphorus, phosphorus(III) oxide, P_4O_6, and phosphorus(V) oxide, P_4O_{10}; they give rise to phosphorous and phosphoric acids respectively.

(h) There are many phosphorous acids and, except for orthophosphorous acid, H_3PO_3, they are polymeric, containing —P—O—P— bonds. Orthophosphorous acid and acidified solutions of orthophosphites are reducing agents; for example, they will convert iodine solutions to iodide ions. Disodium hydrogen orthophosphite, $(Na^+)_2HPO_3^{2-}$, gives an alkaline reaction in water, i.e. orthophosphorous acid is relatively weak.

(i) There are many phosphoric acids and, except for orthophosphoric acid, H_3PO_4, their structures contain —P—O—P— bonds. Orthophosphoric acid and orthophosphates give a yellow precipitate of ammonium phosphomolybdate when treated with a solution of ammonium molybdate, to which a few drops of concentrated nitric acid have been added. The action of heat on orthophosphoric acid produces a glassy material, which is a mixture of metaphosphoric acids (empirical formula HPO_3) and these have cyclic structures. These acids, and the metaphosphates, also respond to the ammonium molybdate test, but at a slower rate than do orthophosphoric acid and orthophosphates.

(j) Phosphorus forms two chlorides, phosphorus trichloride, PCl_3, and phosphorus pentachloride, PCl_5. The former is a colourless liquid and the latter a white solid. Both are hydrolysed by water to give acidic solutions.

(k) Arsenic, antimony and bismuth form the 3-valent oxides. Arsenic(III) oxide is acidic, antimony(III) oxide is amphoteric and bismuth(III) oxide is basic. Arsenic and antimony also form the 5-valent oxides which are acidic.

(l) There is some doubt about the nature of arsenious acid (formed by the action of water on arsenic(III) oxide) but arsenites exist, for example, sodium arsenite, $(Na^+)_3AsO_3^{3-}$, which act as reducing agents in solution. Antimonous acid is unknown but meta-antimonites exist. Arsenic acid is obtained by the action of water on arsenic(V) oxide and gives rise to orthoarsenates which contain the AsO_4^{3-} ion. The orthoarsenate ion is an oxidising agent in strongly acidic solution; for example, it will liberate iodine from iodide solutions. Antimonic acid is unknown, but solid antimonates such as $K^+[Sb(OH)_6]^-$ exist and, although there is no higher acid of bismuth corresponding to arsenic acid, the impure salt, sodium bismuthate, $Na^+BiO_3^-$, can be made. This salt is a powerful oxidising agent; for example, it will oxidise manganese(II) ions to permanganate ions.

(m) Arsenic and antimony from both the 3 and 5-valent sulphides which dissolve in sodium hydroxide solution, i.e. they show 'acidic' properties. Bismuth only forms the 3-valent sulphide, which is unaffected by treatment with sodium hydroxide solution.

(n) Arsenic trichloride, $AsCl_3$, is a covalent liquid which is hydrolysed by water to hydrated arsenic(III) oxide. Antimony and bismuth trichlorides are solids and incompletely hydrolysed by water to give oxysalts, e.g. SbOCl.

(o) Bismuth and to a lesser extent antimony form some salts containing the M^{3+} ion. Typical salts of bismuth are the nitrate, $Bi^{3+}(NO_3^-)_3$, and the basic carbonate, $(BiO^+)_2CO_3^{2-}$.

7

OXIDATION-REDUCTION— SOME REDOX REACTIONS

7.1 Early Definition of Oxidation and Reduction

The simple definition of oxidation as addition of oxygen or removal of hydrogen, and reduction as addition of hydrogen or removal of oxygen has frequently been used to interpret chemical reactions. The two processes are complementary; no oxidation process can take place without a corresponding reduction.

7.2 Extension of Early Ideas

For reactions in which ionic compounds participate, the scope of oxidation-reduction is extended and oxidation is defined as loss of electrons and reduction as gain of electrons. The following experiment links this definition of redox with the early simple definition.

Action of hydrogen sulphide on a solution of iron(III) chloride

Caution: do this experiment in a fume cupboard.

Place about 5 cm^3 of an approximately 0·1M solution of iron(III) chloride in a test-tube and then pass a stream of hydrogen sulphide through it for several seconds. Notice the formation of a fine precipitate which is sulphur. Filter the resulting mixture (some sulphur may pass through the filter paper) and add approximately 2M sodium hydroxide solution to about 2 cm^3 of the filtrate. If the precipitate is brown after adding the sodium hydroxide solution, pass more hydrogen sulphide through the rest of the filtrate and test again with sodium hydroxide solution. What colour is the hydroxide precipitate? What has happened to the iron(III) ions in this reaction?

An explanation of this reaction can be given in terms of ions. Sulphur is formed in the reaction and the iron(III) chloride is converted into iron(II) chloride thus:

$$\underset{\underset{\text{reduction}}{\underbrace{}}}{\overset{\overset{\text{oxidation}}{\overbrace{}}}{H_2S + 2FeCl_3 \rightarrow S + 2FeCl_2 + 2HCl}}$$

The conversion of hydrogen sulphide into sulphur is oxidation by loss of hydrogen, so since oxidation and reduction are complementary processes the iron(III) chloride is reduced to iron(II) chloride. In terms of ions the reaction is:

$$H_2S + 2Fe^{3+}(aq) \rightarrow S + 2Fe^{2+}(aq) + 2H^+(aq)$$

Thus the iron(III) ions are reduced to iron(II) ions by gain of electrons; likewise the hydrogen sulphide is oxidised to sulphur and hydrogen ions by the loss of electrons. This can be seen more clearly by writing the partial equations:

$$2Fe^{3+}(aq) + 2e^- \rightarrow 2Fe^{2+}(aq) \qquad \text{oxidation}$$
$$H_2S \rightarrow S + 2H^+(aq) + 2e^- \qquad \text{reduction}$$

Similarly the displacement of copper from copper(II) sulphate solution by zinc can be expressed by an ionic equation:

$$Zn + Cu^{2+}(aq) \rightarrow Zn^{2+}(aq) + Cu$$

or in terms of partial equations:

$$Zn \rightarrow Zn^{2+}(aq) + 2e^- \qquad \text{oxidation}$$
$$Cu^{2+}(aq) + 2e^- \rightarrow Cu \qquad \text{reduction}$$

7.3 Experimental Evidence for Electron Transfer

All ionic reactions which involve the loss and gain of electrons can be performed under conditions in which an electric current is generated; such an arrangement is known as an electrochemical cell.

Reaction between solutions of iron(III) chloride and potassium iodide

Set up the apparatus as shown in Fig. 7.1 using potassium iodide

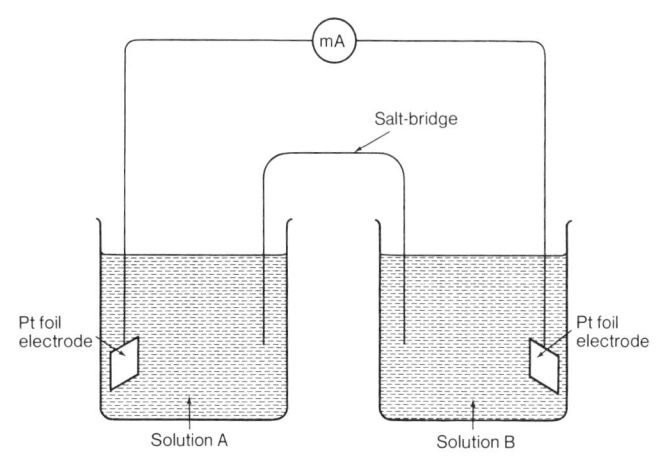

Fig. 7.1
The reaction between ionic substances in solution can be made to generate an electric current.

solution (solution A) and iron(III) chloride solution (solution B). The salt-bridge consists of a piece of filter paper previously soaked in a saturated solution of potassium nitrate. Read the current flowing on the milliammeter (full scale deflection 1 mA); test for iodine in solution A by adding starch solution.

The overall reaction is:

$$2Fe^{3+}(aq) + 2I^-(aq) \rightarrow 2Fe^{2+}(aq) + I_2$$

Write the two partial equations for this reaction, noting the oxidation, the reduction, the oxidising agent and the reducing agent. In which direction does electron flow take place round the external circuit? (i.e. from left to right, or from right to left?).

7.4 Redox Reactions Involving Electron Transfer and either Bond Breaking or Bond Making

(a) Reaction between sodium thiosulphate solution and iodine solution

Using the apparatus shown in Fig. 7.1, place a solution of sodium thiosulphate in one beaker (solution A) and a solution of iodine in potassium iodide solution in the other beaker (solution B). Again use a potassium nitrate salt-bridge and read the current flowing on the milliammeter.

The overall reaction, which involves the formation of an S—S bond in addition to electron transfer, is:

$$2S_2O_3^{2-}(aq) + I_2 \rightarrow S_4O_6^{2-}(aq) + 2I^-(aq)$$

Write two partial equations for this reaction, noting the oxidation, reduction, the oxidising agent and the reducing agent. In which direction does electron flow take place through the external circuit?

(b) Reaction between an acidified solution of potassium permanganate and a solution of iron(II) sulphate solution

Using the apparatus shown in Fig. 7.1, place a solution of iron(II) sulphate in one beaker (solution A) and a solution of potassium permanganate (previously acidified with some approximately M sulphuric acid) in the other beaker (solution B). Use a potassium nitrate salt-bridge to complete the circuit and read the current flowing.

The overall reaction, which involves the breaking of manganese–oxygen bonds in addition to electron transfer is:

$$MnO_4^-(aq) + 8H^+(aq) + 5Fe^{2+}(aq) \rightarrow 5Fe^{3+}(aq) + Mn^{2+}(aq) + 4H_2O$$

Again construct partial equations to show the oxidation and reduction reactions. What is the purpose of the dilute sulphuric acid which is used to acidify the potassium permanganate solution?

7.5 Redox is Relative

In terms of electron transfer, oxidation is the loss of electrons and reduction the gain of electrons; thus the substance which is oxidised, i.e. loses electrons, is the reducing agent. Similarly the substance which is reduced, i.e. gains electrons, is the oxidising agent. Substances which behave as oxidising agents under one set of conditions can be forced to assume the rôle of reducing agents, if they react with stronger oxidising agents than themselves; thus sulphur dioxide will oxidise burning magnesium (experiment 8.11(d)(i), p. 120), but moist sulphur dioxide (sulphurous acid) reduces a variety of substances (Experiments 8.12(a)(i)(ii)(iii)(iv)(v), p. 121). Similarly hydrogen peroxide solution will oxidise a variety of substances (Experiment 8.6(e)(ii), p. 112), but the stronger oxidising agent potassium permanganate forces it to assume the role of reducing agent (Experiment 8.6(e)(iii), p. 112).

7.6 Quantitative Treatment of Redox Systems in Aqueous Solution

When a metal is placed in a solution of its ions, a potential difference is set up between the metal and the solution. The simple model of a metal as an assembly of positive ions held together by a kind of 'electron glue' is useful in visualising what has happened. There is a tendency for metal ions to leave the metal lattice and go into solution, thus leaving an excess of electrons and hence a negative charge on the metal; there is also a reverse tendency for metal ions from the solution to deposit on the metal leading to a positive charge on the metal. In practice one effect is greater than the other, so a potential difference is set up between the metal and the solution. The value of this potential difference for a particular metal depends upon the concentration of metal ions and the temperature, and is called an electrode potential.

It is not possible to measure electrode potentials absolutely, since the very act of carrying out a measurement would necessitate the introduction of another metal into the solution which would set up its own electrode potential. Electrode potentials therefore have to be measured against some reference standard, and the one adopted is the hydrogen electrode. This consists of hydrogen gas at one atmosphere pressure in contact with a solution of hydrogen ions of concentration 1 mol dm^{-3} at 298K; a platinum electrode coated with platinum black is incorporated into the system to catalyse the attainment of equilibrium between the hydrogen gas and the hydrogen ions in solution (Fig. 7.2, p. 100). The electrode potential of the hydrogen electrode under these standard conditions is called a **standard electrode potential** and is arbitrarily assigned a value of zero.

The **standard electrode potential** for the system M^{n+}/M is found by dipping the metal M into a solution of metal ions, M^{n+}, of concentration 1 mol dm^{-3}, and connecting it to a standard hydrogen electrode via a

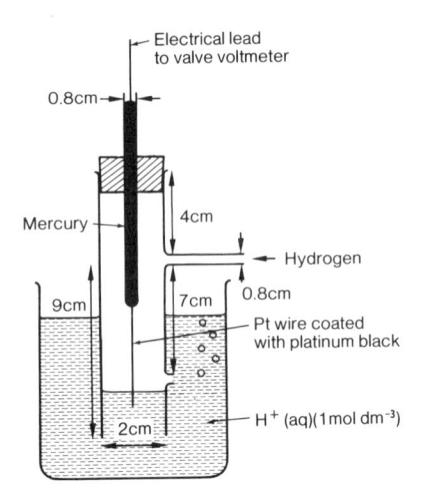

Fig. 7.2
The standard hydrogen electrode.

potassium chloride or potassium nitrate salt-bridge, the whole assembly being at a temperature of 298K. The potential difference developed is read either on a previously calibrated potentiometer or on a high resistance voltmeter, e.g. a valve voltmeter (so that no current flows) (Fig. 7.3). The negative pole of the cell is allotted a negative electrode potential (this is the convention recommended by the International Union of Pure and Applied Chemistry—IUPAC—and is the one used here).

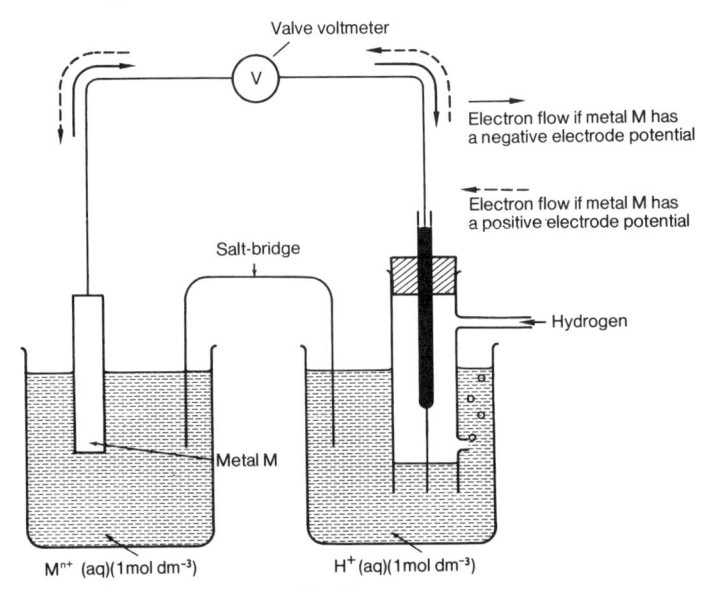

Fig. 7.3
Apparatus for measuring standard electrode potentials.

Copper in contact with copper(II) ions of concentration 1 mol dm^{-3} has a positive standard electrode potential, while zinc in contact with zinc ions of concentration 1 mol dm^{-3} has a negative standard electrode potential (Fig. 7.4).

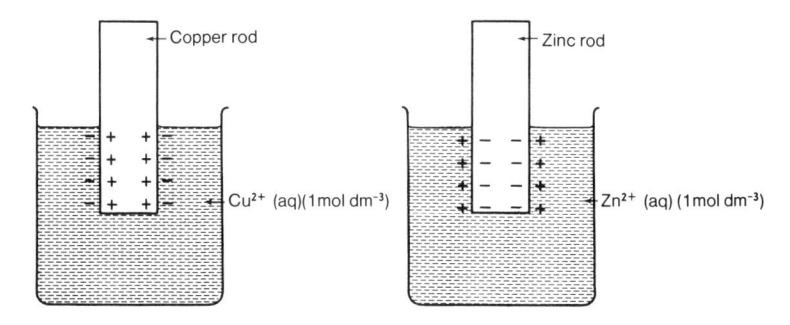

Fig. 7.4
Copper has a positive standard electrode potential and zinc has a negative standard electrode potential.

Determination of the standard electrode potentials of copper and zinc

Set up the apparatus shown in Fig. 7.3 using a strip of copper foil, previously cleaned with emery paper, dipping into a solution of copper(II) sulphate of concentration 1 mol dm^{-3}. A hydrogen electrode, of dimensions shown in Fig. 7.2 is convenient, and this is joined to the copper electrode assembly via a piece of filter paper strip, previously dipped into a saturated solution of potassium nitrate. Switch on the valve voltmeter and allow several minutes for the valves to warm up fully. Zero the instrument and, having selected the appropriate D.C. voltage scale, connect it to the electrochemical cell assembly. Pass a slow stream of hydrogen gas, preferably from a cylinder, through the hydrogen electrode (use hydrochloric acid of concentration 1 mol dm^{-3} as the electrolyte). Record the voltage registered on the valve voltmeter; if the voltmeter reading shows a deflection in the wrong direction you will have to change over the leads, alternatively there may be a switch on the instrument itself for correcting this. Record the polarity of the copper electrode with respect to the hydrogen electrode and quote the standard electrode potential of copper ($E^{\ominus}_{Cu}$) as + or − so many volts.

Repeat the experiment but this time replace the copper electrode assembly with a clean piece of zinc foil dipping into a solution of zinc sulphate of concentration 1 mol dm^{-3}. Record the value and sign of the standard electrode potential of zinc ($E^{\ominus}_{Zn}$).

Decide whether the process:

$$Cu \rightarrow Cu^{2+}(aq) + 2e^{-}$$

is more or less difficult than the process:

$$Zn \rightarrow Zn^{2+}(aq) + 2e^{-}$$

What would be the e.m.f. of the cell constructed from a copper electrode and a zinc electrode assembly? Check your answer by joining the two electrode assemblies together via a potassium nitrate salt-bridge. Write an ionic equation for the chemical reaction that takes place in this electrochemical cell.

7.7 Use of Electrode (Redox) Potentials

A number of standard electrode (redox) potentials are given in Table 7A.

TABLE 7A   Some Standard Electrode (Redox) Potentials

Reaction	$E^{\ominus}$/Volts
$K^+ + e^- \rightarrow K$	-2.92
$Ca^{2+} + 2e^- \rightarrow Ca$	-2.87
$Na^+ + e^- \rightarrow Na$	-2.71
$Mg^{2+} + 2e^- \rightarrow Mg$	-2.37
$Zn^{2+} + 2e^- \rightarrow Zn$	-0.76
$Fe^{2+} + 2e^- \rightarrow Fe$	-0.44
$2H^+ + 2e^- \rightarrow H_2$	0.00
$S + 2H^+ + 2e^- \rightarrow H_2S$	$+0.14$
$Cu^{2+} + 2e^- \rightarrow Cu$	$+0.34$
$I_2 + 2e^- \rightarrow 2I^-$	$+0.54$
$Fe^{3+} + e^- \rightarrow Fe^{2+}$	$+0.76$
$Ag^+ + e^- \rightarrow Ag$	$+0.80$
$Br_2 + 2e^- \rightarrow 2Br^-$	$+1.07$
$Cr_2O_7^{2-} + 14H^+ + 6e^- \rightarrow 2Cr^{3+} + 7H_2O$	$+1.33$
$Cl_2 + 2e^- \rightarrow 2Cl^-$	$+1.36$
$MnO_4^- + 8H^+ + 5e^- \rightarrow Mn^{2+} + 4H_2O$	$+1.52$
$S_2O_8^{2-} + 2e^- \rightarrow 2SO_4^{2-}$	$+2.01$

(Increasingly powerful oxidising agents ← / → Increasingly powerful reducing agents)

Standard redox potentials for non-metals that produce negative ions in aqueous solution can be determined in a similar manner to that used for metals. Thus the standard redox potential of chlorine can be determined using an electrode consisting of chlorine gas at one atmosphere pressure in equilibrium with an aqueous solution of chloride ions of concentration 1 mol dm^{-3} and into which a platinum wire is dipping. Similarly for metals of variable valency, redox potentials for one ion in equilibrium with another of different charge can be determined. The standard redox potential for the system $Fe^{3+}(aq)/Fe^{2+}(aq)$ is obtained by coupling to a standard hydrogen electrode a half-cell containing a solution which is of concentration 1 mol dm^{-3} with respect to both iron(II) and iron(III) ions and into which a platinum wire is dipping. The function of the platinum wire is that of a conductor and catalyst.

The half-cell for the system $MnO_4^-(aq)/Mn^{2+}(aq)$ consists of a solution which is of concentration 1 mol dm^{-3} with respect to permanganate ions, manganese(II) ions and also hydrogen ions; again, a platinum wire is immersed in the solution.

Standard redox potentials may be used to predict possible ionic reactions in aqueous solution. For example, magnesium with a standard redox potential of -2.37 V should reduce an aqueous solution of copper(II) ions to copper and this indeed does take place in practice. We can depict this system as follows:

$$\text{Electron flow} \longrightarrow$$

$$\underset{-\text{ve}}{\text{Mg}} \, | \, \text{Mg}^{2+}(\text{aq})(1 \, \text{mol} \, \text{dm}^{-3}) \, \vdots \, \text{Cu}^{2+}(\text{aq})(1 \, \text{mol} \, \text{dm}^{-3}) \, | \, \underset{+\text{ve}}{\text{Cu}}$$

(dotted line denotes a salt-bridge)

the e.m.f. of the cell ($E^\ominus_{\text{total}}$) is defined to be the standard redox potential of the right-hand electrode minus the standard redox potential of the left-hand electrode.

$$E^\ominus_{\text{total}} = E^\ominus_{\text{R.H.S}} - E^\ominus_{\text{L.H.S}} = +0.34 - (-2.37) = +2.71 \, \text{V}$$

Using this convention, a positive e.m.f. means that the left-hand electrode (magnesium) is potentially capable of reducing copper(II) ions to copper, i.e.

$$\text{Mg} + \text{Cu}^{2+}(\text{aq}) \rightarrow \text{Mg}^{2+}(\text{aq}) + \text{Cu}$$

Similarly, manganese(II) ions should be capable of reducing persulphate ions to sulphate ions in aqueous solution, the manganese(II) ions being converted into permanganate ions (see Experiment 8.18(b)(iv), p. 129). This system can be indicated as below:

$$\text{Electron flow} \longrightarrow$$

$$\underset{-\text{ve}}{\text{Pt}} \, | \, \text{Mn}^{2+}(\text{aq}), \text{H}^+(\text{aq}), \text{MnO}_4^-(\text{aq}) \, \vdots \, \text{S}_2\text{O}_8^{2-}(\text{aq}), \text{SO}_4^{2-}(\text{aq}) \, | \, \underset{+\text{ve}}{\text{Pt}}$$

$$E^\ominus_{\text{total}} = +2.01 - (+1.52) = +0.49 \, \text{V}$$

The complete equation for this reaction may be obtained by subtraction of the two partial equations:

$$5\text{S}_2\text{O}_8^{2-}(\text{aq}) + 10\text{e}^- \rightarrow 10\text{SO}_4^{2-}(\text{aq})$$
$$2\text{MnO}_4^-(\text{aq}) + 16\text{H}^+(\text{aq}) + 10\text{e}^- \rightarrow 2\text{Mn}^{2+}(\text{aq}) + 8\text{H}_2\text{O}$$

$$2\text{Mn}^{2+}(\text{aq}) + 8\text{H}_2\text{O} + 5\text{S}_2\text{O}_8^{2-}(\text{aq}) \rightarrow$$
$$\rightarrow 2\text{MnO}_4^-(\text{aq}) + 16\text{H}^+(\text{aq}) + 10\text{SO}_4^{2-}(\text{aq})$$

Although values of standard redox potentials allow one to make predictions about chemical reactions that may be carried out in aqueous solution, they give no information as to how fast a particular reaction is likely to proceed. Indeed, it may well be that a reaction predicted to take place does not actually do so, simply because the rate is so very slow. In fact, the reaction between manganese(II) ions and persulphate ions in acid solution has to be catalysed by silver ions (see Experiment 8.18(b)(iv), p. 129). In addition, it should be mentioned that standard redox potentials refer to solutions of concentration 1 mol dm^{-3} and the removal of an ion by precipitation, for example, would obviously upset things. It is thus necessary to use values of standard redox potentials with some caution.

7.8 Variation of Redox Potentials with Concentration

If a metal is immersed in an aqueous solution of its ions and the concentration of metal ions is decreased, then there will be less tendency for the metal ions to deposit on the metal, i.e. the electrode potential of the metal will become less positive. The way in which the electrode potential of a metal depends on the metal ion concentration is shown in the following experiment.

(a) The electrode potential of silver at varying silver ion concentrations

Set up the apparatus shown in Fig. 7.5 which consists of a clean piece of copper dipping into M copper(II) sulphate solution and a clean piece of silver dipping into 0·01M silver nitrate solution. Join the two solutions

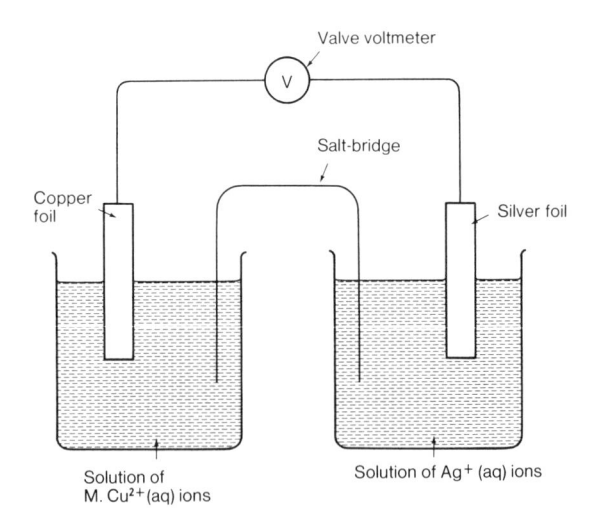

Fig. 7.5
The electrode potential of silver at varying silver ion concentrations.

by a salt-bridge (a piece of filter paper previously dipped into a saturated solution of potassium nitrate). Switch on the valve voltmeter and allow several minutes for it to warm up, select the appropriate D.C. voltage scale and zero the instrument. Now connect it to the cell, the positive terminal to the silver electrode, and measure the e.m.f. of the cell. Repeat the experiment, but vary the silver ion concentration (0·0033M, 0·001M, 0·00033M and 0·0001M).

The e.m.f. of the cell is given by the equation:

$$E_{total} = E_{Ag} - E^{\ominus}_{Cu}$$

$$E_{Ag} = E_{total} + E^{\ominus}_{Cu}$$

$$= E_{total} + 0·34$$

Determine the values of E_{Ag} and then plot a graph of these values (along the y-axis) against $\log_{10}[Ag^+(aq)]$ (these will be negative values). You should find that the graph is a straight line, showing that the electrode potential of silver varies in a logarithmic manner with the silver ion concentration. Deduce the standard electrode potential of silver ($E^{\ominus}_{Ag}$) from the graph and compare this value with that given in Table 7A. Consult a textbook for the Nernst equation—the equation which shows how the electrode potential of a metal is related to the metal ion concentration, and how it is influenced by temperature changes.

(b) The equilibrium constant for the reaction

$$Cu + 2Ag^+(aq) \rightleftharpoons Cu^{2+}(aq) + 2Ag$$

Equilibrium is established when the e.m.f. of the cell is zero, the equilibrium constant for the reaction being:

$$K = \frac{[Cu^{2+}(aq)]_{equilibrium}}{[Ag^+(aq)]^2_{equilibrium}} \quad \text{(units dm}^3\text{ mol}^{-1}\text{)}$$

where the concentrations of the ions are the equilibrium concentrations, i.e. the concentrations of the ions when the e.m.f. of the cell is zero.

Extrapolate the graph until the line cuts the x-axis and read off $\log_{10}[Ag^+(aq)]$ for $E_{total} = 0$. The $Cu^{2+}(aq)$ concentration is 1 mol dm^{-3}. Convert the logarithm of the silver ion concentration to silver ion concentration, and hence deduce the equilibrium constant for this reaction.

8

GROUP 6B OXYGEN, SULPHUR, SELENIUM, TELLURIUM AND POLONIUM

8.1 Some Physical Data of Group 6B Elements

	Atomic number	Electronic configuration	Density/ $g\,cm^{-3}$	M.p./ K	B.p./ K	Atomic radius/nm	Ionic radius/nm X^{2-}
O	8	2.6 $1s^22s^22p^4$		54	90	0·074	0·140
S	16	2.8.6 $...2s^22p^63s^23p^4$	2·07 (rhombic)	392	718	0·104	0·184
Se	34	2.8.18.6 $...3s^23p^63d^{10}4s^24p^4$	4·80 (grey)	490	958	0·117	0·198
Te	52	2.8.18.18.6 $...4s^24p^64d^{10}5s^25p^4$	6·25	723	1663	0·137	0·221
Po	84	2.8.18.32.18.6 $...5s^25p^65d^{10}6s^26p^4$				0·152	

8.2 Some General Remarks about Group 6B

The Group 6B elements show the usual gradation from non-metallic to metallic properties with increasing atomic number that occurs in any Periodic Group. Oxygen and sulphur are non-metals, selenium and tellurium are semiconductors and polonium is metallic.

These elements can enter into chemical combination and complete their octets by gaining two electrons to form the 2-valent ion, e.g. O^{2-}, S^{2-}, except for polonium which is too metallic, and by forming two covalent bonds; for example, the hydrides H_2O, H_2S, H_2Se, H_2Te and H_2Po.

The two heavier members of this Group can form the 4-valent cation X^{4+} (the inert pair effect); for example, there is evidence of the presence of Te^{4+} ions in the dioxide $Te^{4+}(O^{2-})_2$ and of Po^{4+} ions in the dioxide, $Po^{4+}(O^{2-})_2$, and sulphate, $Po^{4+}(SO_4^{2-})_2$.

Because sulphur, selenium, tellurium and polonium have vacant d orbitals that can be utilised without too great an energy change, they are able to form covalent compounds in which the octet of electrons is expanded; for instance, the valencies of sulphur in H_2S, SCl_4 and SF_6 are two, four and six respectively. Oxygen, in common with other first row members of the Periodic Table, cannot expand its octet.

Oxygen exists in the form of discrete molecules, a double bond uniting two oxygen atoms together, O=O. The atoms of the other Group 6B elements do not multiple bond to themselves and sulphur, in particular, shows a strong tendency to catenate, puckered S_8 rings being present in rhombic and monoclinic sulphur (the two main allotropic forms of this element). There are two forms of selenium corresponding in structure to rhombic and monoclinic sulphur in which Se_8 rings are present. These forms, however, are readily converted into a 'metallic' form of the element called grey selenium which contains long spiral chains of selenium atoms. As far as is known, there is only one form of tellurium which has the same structure as grey selenium. Polonium is truly metallic, exhibiting a co-ordination number of six in both its allotropic forms.

Since the chemistry of oxygen and sulphur have little in common, these two elements are treated separately. No experimental work on selenium, tellurium and polonium is included; indeed polonium is radioactive and is only available in milligramme quantities, but such is the sophistication of modern chemical techniques that a significant amount of chemistry has been done on this very small scale.

8.3 Extraction of Oxygen and Sulphur

Oxygen occurs in the atmosphere to the extent of about 21% by volume and is extracted from the air by first liquefying it and then fractionally distilling the liquid air.

Sulphur occurs in combination with many metals as sulphides and sulphates. It occurs in the free state in Japan, and in Texas and Louisiana, where it was discovered by Frasch. It is from this latter source that most of the world's sulphur is derived (see an inorganic textbook for a description of the Frasch process).

OXYGEN

8.4 Reactions of Oxygen

Oxygen is very reactive and forms compounds with all other elements except the noble gases and, apart from the halogens and some unreactive metals, these can be made to combine directly with oxygen under the right conditions.

8.5 Some Methods of Forming Oxygen

(a) By the decomposition of higher oxides

Place about 0·3 g of dilead(II) lead(IV) oxide in an ignition tube and heat. Test for oxygen with a glowing splint.
Repeat the experiment with lead(IV) oxide and manganese(IV) oxide.

(b) By the decomposition of peroxides

(i) *Catalytic decomposition of hydrogen peroxide, H_2O_2*

Place about 2 cm^3 of '20 volume' hydrogen peroxide solution in a test-tube and add a pinch of manganese(IV) oxide. See if the tube becomes hot and test for oxygen with a glowing splint.

(ii) *Action of heat on barium peroxide, $Ba^{2+}O_2^{2-}$*

Carry out Experiment 3.5(c) p. 33, if you have not done so previously.

(c) By the thermal decomposition of some oxysalts

Heat about 0·3 g of the following salts in separate ignition tubes and test for the evolution of oxygen: potassium nitrate, $K^+NO_3^-$, potassium permanganate, $K^+MnO_4^-$, potassium persulphate, $(K^+)_2S_2O_8^{2-}$, potassium bromate(V), $K^+BrO_3^-$, potassium chlorate(V), $K^+ClO_3^-$, potassium iodate(V), $K^+IO_3^-$. Consult an inorganic textbook for the products of the above reactions.

8.6 The Hydrides of Oxygen

Oxygen forms two hydrides, namely water, H_2O, and hydrogen peroxide, H_2O_2. The latter compound is unstable with respect to water and molecular oxygen (see Experiment 8.5(b)(i)).

(a) The polar nature of the water molecule

The water molecule has an angular structure (Fig. 8.1) the two bonding pairs and two lone pairs of electrons being arranged approximately tetrahedrally round the oxygen atom (to minimise repulsion between electron pairs). The molecule is also extensively polarised because of the high electronegativity of the oxygen atom, partial positive charges residing on each hydrogen atom, and a negative charge on the oxygen.

Deflection of water with a charged rod

Fill a burette with distilled water and then allow it to flow out of the jet into an empty beaker. Bring a charged rod (a perspex rod rubbed with cloth is suitable) up to the stream of water and note that the water is deflected.

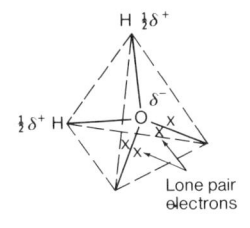

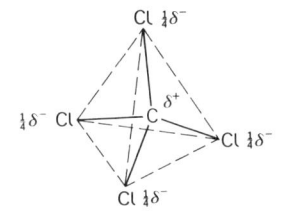

Fig. 8.1
The structure of the water molecule.

Fig. 8.2
The structure of the tetrachloromethane molecule.

The molecule of tetrachloromethane is tetrahedral (see Fig. 8.2) with partial negative charges on each chlorine atom and a partial positive charge on the carbon. Would you expect tetrachloromethane to be deflected by a charged rod? Carry out the experiment and explain the result.

(b) The solvent action of water

Many electrovalent compounds are soluble in water, the separated ions being surrounded by water molecules (a process called hydration). Thus consider the dissolution of an electrolyte A^+B^- in water which is itself polarised in the sense:

$$H^{\frac{1}{2}\delta+} \diagdown O^{\delta-} \diagup H^{\frac{1}{2}\delta+}$$

The negative end of the water dipole points towards the cation A^+, and the positive end towards the anion B^-. Hydration is an exothermic process (chemical bonds are formed), the energy released being known as hydration energy. This energy release helps to offset the large lattice energy that must be absorbed in the process of separating the ions in a crystal.

It is because water is appreciably polarised that it is such a good solvent for many ionic substances.

Action of water and tetrachloromethane on some ionic substances

Show that substances like anhydrous copper(II) sulphate and potassium chloride dissolve readily in cold water. Now try the effect of a little tetrachloromethane on these two salts. Do they dissolve in this liquid? Explain your result.

(c) Salt hydrolysis

An aqueous solution of disodium hydrogen orthophosphate, $(Na^+)_2HPO_4^{2-}$, shows an alkaline reaction to indicators (owing to the presence of OH⁻ ions). In solution, sodium and hydrogen orthophosphate ions are present, the latter being strong proton acceptors; the alkalinity of the solution is thus due to the reaction:

$$\underset{\text{Acid(1)}}{H_2O} + \underset{\text{Base(2)}}{HPO_4^{2-}} \rightleftharpoons \underset{\text{Base(1)}}{OH^-} + \underset{\text{Acid(2)}}{H_2PO_4^-}$$

The highly charged small cations of metals are strongly hydrated in aqueous solution. Furthermore, the small highly charged cations exert a considerable attraction on the oxygen atoms of water molecules, thereby weakening the links between the hydrogen and oxygen atoms. Under these conditions solvent water molecules are able to act as a base and thus give rise to an acid solution. An aqueous solution of iron(III) chloride, for instance, shows an acid reaction:

$$\underset{\text{Acid(1)}}{[Fe(H_2O)_6]^{3+}} + \underset{\text{Base(2)}}{H_2O} \rightleftharpoons \underset{\text{Base(1)}}{[Fe(H_2O)_5(OH)]^{2+}} + \underset{\text{Acid(2)}}{H_3O^+}$$

Action of water on some salts

Make up dilute solutions of the following salts and try the effect of a few drops of Universal Indicator solution on these solutions: sodium sulphate, sodium carbonate, sodium hydrogen carbonate, copper(II) sulphate, iron(II) sulphate (for the last two use pH paper). List those salts that give an alkaline reaction and those that give an acidic reaction in water. Do any show a neutral reaction?

Differentiate between the reactions of sodium carbonate and sodium hydrogen carbonate solutions by seeing what effect these two solutions have on a few drops of phenolphthalein (colour change from colourless to red over the pH range 8·3–10). Attempt to explain this result.

From your results, what can you say about the strength of the sulphate ion as a base in aqueous solution and the strength of the hydrated sodium ion as an acid? How would you expect potassium nitrate, nickel nitrate and potassium carbonate to behave in water? Check your predictions.

(d) Purification of water by ion-exchange

Hardness in water is mainly caused by the hydrogen carbonates and sulphates of calcium and magnesium. These salts may be removed by ion-exchange; this process removes ions from impure water and for the purposes of demonstration the two coloured compounds copper(II) sulphate and potassium chromate are used. In the following experiment the ion-exchange materials used are Amberlite I.R.-120 (acidic resin) and Amberlite I.R.A.-410 (basic resin); both should be washed and regenerated according to the manufacturers instructions.

The action of ion-exchange resins **(Demonstration)**

Fill two glass tubes respectively with the acidic resin and the basic resin, glass wool being used to prevent any of these resins entering the exit tubes. Now run a dilute solution of copper(II) sulphate through the acidic resin and note the colour of the effluent. Run the dilute solution of potassium chromate through the basic resin and again note the colour of the effluent. In each case wash the solutions through with some distilled water.

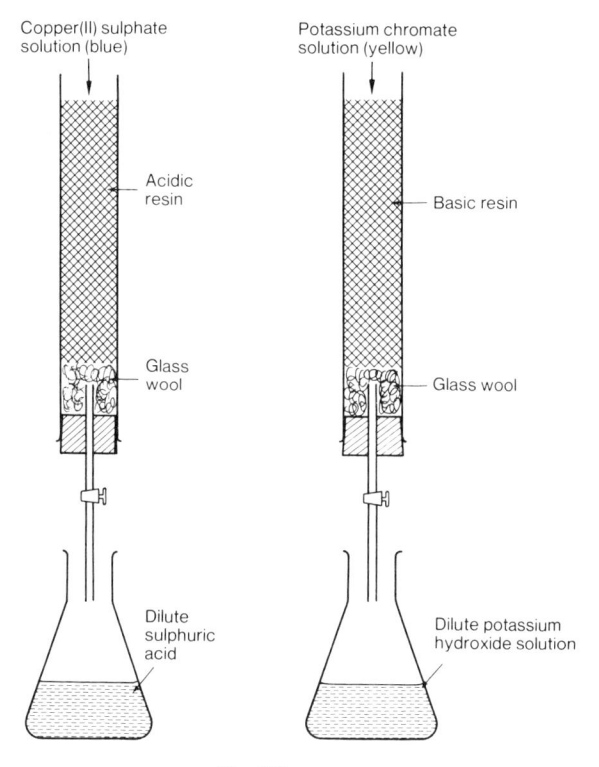

Fig. 8.3
The action of ion-exchange resins.

The action of the acidic resin can be represented thus:

$$Cu^{2+} + SO_4^{2-} + \underset{\text{acidic resin}}{2H-R} \rightarrow \underset{\text{effluent}}{2H^+ + SO_4^{2-}} + \underset{\text{spent resin}}{Cu(R)_2}$$

What do you think would happen if the effluent from the acidic resin were run through a fresh sample of the basic resin? Verify your prediction.

$$\underset{\substack{\text{effluent from} \\ \text{acidic resin}}}{2H^+ + SO_4^{2-}} + \underset{\text{basic resin}}{2R-OH} \rightarrow ? + ?$$

(e) Some reactions of hydrogen peroxide, H_2O_2

(i) *Catalytic decomposition of hydrogen peroxide*

Carry out Experiment 8.5(b)(i) if you have not done it previously:

$$2H_2O_2 \rightarrow 2H_2O + O_2$$

(ii) *Oxidising action of hydrogen peroxide*

Hydrogen peroxide in acidic solution is a strong oxidising agent as the following standard redox potential indicates:

$$H_2O_2 + 2H^+ + 2e^- \rightarrow 2H_2O \qquad E^\ominus = +1.77 \text{ V}$$

It should oxidise aqueous solutions of iron(II) ions to iron(III) ions and iodide ions to iodine as the following standard redox potentials indicate:

$$Fe^{3+} + e^- \rightarrow Fe^{2+} \qquad E^\ominus = +0.76 \text{ V}$$
$$I_2 + 2e^- \rightarrow 2I^- \qquad E^\ominus = +0.54 \text{ V}$$

Place about 2 cm³ of a dilute solution of iron(II) sulphate in a test-tube and acidify it with about the same volume of approximately M sulphuric acid. Now add about the same volume of '20 volume' hydrogen peroxide solution and warm for a few minutes. Cool the solution and then test for the presence of iron(III) ions with sodium hydroxide solution.

$$2Fe^{2+} + H_2O_2 + 2H^+ \rightarrow 2Fe^{3+} + 2H_2O$$

Place about 2 cm³ of a dilute solution of potassium iodide in a test-tube and acidify it with about the same volume of approximately M sulphuric acid. Now add about the same volume of '20 volume' hydrogen peroxide solution and note the result.

$$2I^- + H_2O_2 + 2H^+ \rightarrow I_2 + 2H_2O$$

(iii) *Reducing action of hydrogen peroxide*

The strong oxidising agent potassium permanganate forces hydrogen peroxide to assume the rôle of a reducing agent.

Place about 2 cm³ of a previously acidified solution of potassium permanganate in a test-tube and then add, drop by drop, a '20 volume' hydrogen peroxide solution with shaking. Note the result and explain if it is consistent with the following standard redox potentials:

$$MnO_4^- + 8H^+ + 5e^- \rightarrow Mn^{2+} + 4H_2O \qquad E^\ominus = +1.52 \text{ V}$$
$$O_2 + 2H^+ + 2e^- \rightarrow H_2O_2 \qquad E^\ominus = +0.68 \text{ V}$$

(iv) *Test for hydrogen peroxide*

Place about 2 cm³ of an acidified solution of potassium dichromate in a test-tube and then add about 2 cm³ of diethyl ether (ethoxyethane) **(caution: it is very flammable)**. Now add about 2 cm³ of '20 volume' hydrogen peroxide solution and note the formation of a blue colour in the ether layer. This blue colour is due to chromium peroxide, CrO_5, which is stabilised by complexing with the ether.

SULPHUR

8.7 Reactions of Sulphur

Sulphur combines with most metals when heated. Non-metals that combine directly with sulphur include fluorine, chlorine, oxygen and carbon; hydrogen combines reversibly to a slight extent when passed through molten sulphur near its boiling point.

Sulphur is oxidised by concentrated nitric and sulphuric acids and with hot concentrated solutions of alkalis it forms polysulphides and a thiosulphate (note the tendency of the sulphur atom to catenate):

$$3S + 6OH^- \rightarrow 2S^{2-} + SO_3^{2-} + 3H_2O$$

followed by:

$$S^{2-} + nS \rightarrow S_{n+1}^{2-}$$
polysulphide ion

$$SO_3^{2-} + S \rightarrow S_2O_3^{2-}$$
thiosulphate ion

(a) Reaction of sulphur with iron

Place a small amount of an intimate mixture of iron filings and sulphur in an ignition tube and heat it. Some of the sulphur will vaporise, but go on heating it until a red glow begins to spread through the mixture. Remove the ignition tube from the Bunsen flame and note that the mixture continues to glow. Allow the ignition tube to cool, crack it and extract the lump of iron(II) sulphide.

(b) Reaction of sulphur with chlorine (Demonstration)

Carry out this reaction in a fume cupboard.

Place about 1 g of powdered roll sulphur in the combustion tube (see Fig. 8.4) and apply sufficient heat just to melt it. Pass chlorine over it and

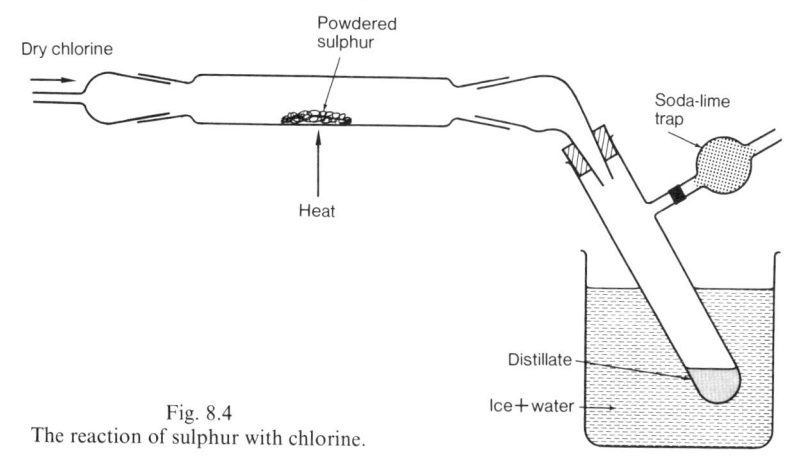

Fig. 8.4
The reaction of sulphur with chlorine.

collect the liquid product in a tube cooled in ice and water. The liquid product is mainly disulphur dichloride, S_2Cl_2. If it is required pure, it may be distilled and the fraction boiling in the region of 138°C collected. Keep the product in a stoppered bottle for a later experiment.

8.8 The Allotropy of Sulphur

Unlike oxygen which is a discrete molecule, two atoms being united by a double bond, sulphur atoms show a marked reluctance to double bond with themselves and the two main allotropes of sulphur contain S_8 molecules, in which single bonds unite sulphur atoms into a puckered octagonal ring. The high relative molecular mass of these S_8 structural units explains why sulphur, unlike oxygen, is a solid.

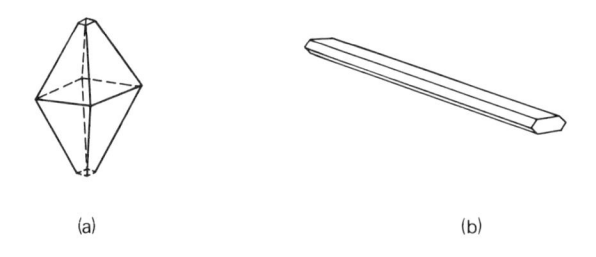

(a)   (b)

Fig. 8.5
The crystal shapes of (a) rhombic sulphur, (b) monoclinic sulphur.

(a) Rhombic sulphur

This is the form of sulphur normally encountered and consists of S_8 structural units packed together to give a crystal whose shape is shown in Fig. 8.5(a). The crystal is yellow, transparent and has a density of $2·06 \, g \, cm^{-3}$.

Preparation of rhombic sulphur **(Demonstration)**

Place about 2 g of powdered roll sulphur in a boiling-tube and add about 10 cm³ of carbon disulphide (**caution: carbon disulphide is evil-smelling and poisonous so carry out this experiment in a fume cupboard; it is also very flammable so keep well away from naked lights**). Shake the tube and then filter the solution into a small conical flask. Cover the conical flask with filter paper, which has a few small holes punched through it, and allow the carbon disulphide to evaporate slowly in the fume cupboard. Examine the crystals which should be present after several hours.

(b) Monoclinic sulphur

If crystallisation of sulphur from a suitable solvent occurs at a temperature in excess of 368·6 K, the crystals take the shape shown in Fig. 8.5(b).

They are light yellow in colour and have a density of $1.96\,\text{g cm}^{-3}$. Like rhombic sulphur, monoclinic sulphur consists of S_8 structural units, but these are arranged differently in the crystal lattice.

Preparation of monoclinic sulphur (**Demonstration**)

Place about 2 g of powdered roll sulphur in a boiling-tube and add about $10\,\text{cm}^3$ of turpentine. Warm to dissolve the sulphur (**caution: turpentine is flammable**). Place the tube in a beaker and insulate the tube with cotton wool, so that the solution cools very slowly ensuring that crystallisation occurs above a temperature of 368·6K. Examine the sulphur crystals the next day.

(c) The action of heat on sulphur and the formation of plastic sulphur

A series of complex changes occur when sulphur is heated. If nearly boiling liquid sulphur is poured into cold water, plastic sulphur is formed. Carry out the next experiment before consulting an inorganic textbook for an explanation of the changes involved.

Formation of plastic sulphur

Put sufficient powdered roll sulphur in a test-tube to fill the tube to a depth of about 4 cm and then slowly heat it, rotating the tube to get even heating. Go on heating until the liquid sulphur formed is almost boiling, and then pour it into a beaker containing cold water. Note all the changes in colour and viscosity of the liquid sulphur, and examine the texture of the plastic sulphur. Roll the plastic sulphur between the fingers and note what happens. Leave it for 24 hours and then examine it again. What do you think has happened?

8.9 The Hydrides of Sulphur

Sulphur forms a number of hydrides which contain catenated sulphur atoms such as H_2S_2, H_2S_3, H_2S_4 etc. They are yellow oils which readily decompose into hydrogen sulphide, H_2S, and free sulphur. Hydrogen sulphide is the only hydride of sulphur of any importance and is the sulphur analogue of water. Can you explain why hydrogen sulphide, H_2S, is a gas whereas water, H_2O, is a liquid?

(a) Formation of hydrogen sulphide, H_2S

Caution: hydrogen sulphide is evil-smelling and very poisonous, so do all the experiments with this gas in a fume cupboard.

Place a small lump of iron(II) sulphide in a test-tube and add about $3\,\text{cm}^3$ of approximately 2M hydrochloric acid. Warm gently and hold a piece of filter paper strip previously dipped into lead(II) ethanoate solution near the mouth of the tube. Record your observation. Write equations for the action of H^+ ions on iron(II) sulphide and for the action of hydrogen sulphide on Pb^{2+} ions.

(b) The acidic properties of hydrogen sulphide

(i) *Action of Universal Indicator solution on hydrogen sulphide solution*

Make a solution of hydrogen sulphide in distilled water (if possible use a cylinder of hydrogen sulphide for this, since hydrogen sulphide prepared from a Kipp's apparatus may contain some hydrochloric acid spray). To about $2\,cm^3$ of this solution add about 3 drops of Universal Indicator solution and compare the colour with that obtained by adding three drops of Universal Indicator solution to the same volume of distilled water. How would you rate the hydrogen sulphide solution as an acid? Write two equations for the action of water on hydrogen sulphide.

(ii) *Action of sodium hydroxide solution on hydrogen sulphide*

Fill a test-tube with hydrogen sulphide by displacement of water and invert the tube in a solution of sodium hydroxide contained in a small dish. Agitate the tube and note what happens. Write two equations for the action of OH^- ions on hydrogen sulphide.

(c) Hydrogen sulphide as a precipitating agent

Hydrogen sulphide is used in qualitative analysis for precipitating the sulphides of many metals; in practice the precipitation is done under controlled conditions so that two groups of sulphides can be distinguished (analysis Groups II and IV, which are in no way related to the Groups 2 and 4 of the Periodic Classification).

Action of hydrogen sulphide on the sulphates of copper(II), nickel and zinc

Pass hydrogen sulphide through about $3\,cm^3$ of an approximately 0·1M solution of copper(II) sulphate and note the result. Repeat with approximately 0·1M solutions of nickel sulphate and zinc sulphate and record your results. Now add approximately 2M aqueous ammonia drop by drop to the tubes which contained the nickel sulphate and the zinc sulphate and note any changes.

Repeat all three experiments again, but this time add about $2\,cm^3$ of approximately 2M hydrochloric acid to each sulphate solution before passing hydrogen sulphide through them. Record your results.

Hydrogen sulphide in water produces some sulphide ions according to the two equations:

$$H_2O + H_2S \rightleftharpoons H_3O^+ + HS^-$$
$$H_2O + HS^- \rightleftharpoons H_3O^+ + S^{2-}$$

What will be the effect on the sulphide ion concentration of added aqueous ammonia (added OH⁻ ions)? Similarly, what will be the effect on the sulphide ion concentration of added dilute hydrochloric acid (added H_3O^+ ions)? Can you list the sulphides of copper, nickel and zinc in increasing order of solubility from your results and the information given above?

(d) Reducing properties of hydrogen sulphide

Hydrogen sulphide reacts with a variety of substances and reduces them, i.e. it is itself oxidised. In some cases this oxidation of hydrogen sulphide is best expressed as the equation:

$$H_2S + [O] \rightarrow H_2O + S$$

where [O] indicates the oxygen contained in the oxidising agent; in other cases, particularly in aqueous solution, the oxidation of hydrogen sulphide is more appropriately described in terms of the loss of electrons:

$$H_2S \rightarrow 2H^+ + S + 2e^-$$

(i) *Reduction of sulphur dioxide*

Place a few drops of distilled water in a boiling-tube and then fill the tube with sulphur dioxide from a cylinder. Pass hydrogen sulphide through the tube and note the result. Attempt to write an equation for the reaction between hydrogen sulphide and sulphur dioxide (the reaction is catalysed by the few drops of water that were initially added).

(ii) *Reduction of acidified solutions of potassium permanganate and potassium dichromate*

Place about 3 cm³ of a dilute solution of potassium permanganate, previously acidified with approximately M sulphuric acid, in a test-tube. Pass hydrogen sulphide through the solution and note the result.

Repeat the experiment, but this time use a very dilute solution of potassium dichromate previously acidified with approximately M sulphuric acid.

Given the following information about these two reactions, complete balanced partial equations (showing electrons lost and gained) and then balanced equations (electrons eliminated from the equations).

$$MnO_4^- + H^+ + e^- \rightarrow Mn^{2+} + H_2O$$
$$Cr_2O_7^{2-} + H^+ + e^- \rightarrow Cr^{3+} + H_2O$$
$$H_2S \rightarrow 2H^+ + S + 2e^-$$

(iii) *Reduction of potassium iodide solution*

Place about 3 cm³ of a solution of iodine in aqueous potassium iodide in a test-tube and then pass hydrogen sulphide into it. Note the result, and then write partial and complete equations as in (ii) above for the reduction of iodide ions.

8.10 Sulphides

Metallic sulphides are less ionic than the corresponding oxides (why?) and only the sulphides of Groups 1A and 2A appear to be essentially ionic, e.g. $(K^+)_2 S^{2-}$. Aqueous solutions of these ionic sulphides react with sulphur on heating to form a number of polysulphides which contain chain polysulphide anions:

$$(K^+)_2 S^{2-} + nS \rightarrow (K^+)_2 S^{2-}_{n+1}$$

By far the greatest proportion of metallic sulphides are covalent and these are insoluble in water, e.g. ZnS. The sulphides of transition elements behave very much like alloys and many show metallic properties, for example, metallic lustre and conductivity.

Many transition elements also form disulphides in which discrete S_2 groups are present; they too are covalent solids. Iron(II) disulphide, FeS_2, is an important source of sulphur.

(a) Action of water on sodium sulphide, $(Na^+)_2 S^{2-}$

Place about 0·3 g of sodium sulphide in a test-tube and add about 3 cm^3 of distilled water. Test the solution with about 3 drops of Universal Indicator solution and note the result.

Make up another solution of sodium sulphide and boil the solution. Hold a piece of filter paper strip previously soaked in lead(II) ethanoate solution near the mouth of the test-tube and note the result. Write two equations for the action of water on the sulphide ion, i.e. show each stage of the hydrolysis reaction. How would you ensure that these reactions go to completion?

(b) Polysulphide formation with a solution of sodium sulphide

Make up a fairly concentrated solution of sodium sulphide in distilled water and to about 5 cm^3 of this solution add a small amount of powdered roll sulphur. Warm the solution and note the result. What do you think would happen if the final solution is acidified with dilute hydrochloric acid? Test your suggestion.

(c) Action of hydrochloric acid on zinc sulphide, ZnS

Place about 0·3 g of zinc sulphide in a test-tube and add about 2 cm^3 of approximately 2M hydrochloric acid. Warm the solution and then hold a piece of filter paper strip previously soaked in lead(II) ethanoate solution near the mouth of the tube. Record the result. Write an equation for the action of hydrochloric acid on zinc sulphide.

Does this result agree with what you discovered in Section 8.9(c). Would you expect copper(II) sulphide to react with dilute hydrochloric acid? Explain your answer.

(e) Action of heat on iron(II) disulphide, FeS_2

Place a few small lumps of iron(II) disulphide in an ignition tube and heat strongly. Record and explain your observations.

8.11 The Oxides of Sulphur

Sulphur forms several oxides, but the only ones of any importance are sulphur dioxide, SO_2, and sulphur trioxide, SO_3.

(a) Formation of sulphur dioxide, SO_2

Place a few crystals of sodium sulphite in a test-tube and then add about 3 cm^3 of approximately 2M hydrochloric acid. Warm gently and **very cautiously** smell the gas evolved. Hold pieces of filter paper strips dipped respectively into potassium dichromate solution (acidified with dilute sulphuric acid) and potassium permanganate solution (acidified with dilute sulphuric acid) near the mouth of the tube. Note any changes that occur. Write an equation for the action of H^+ ions on sulphite ions, SO_3^{2-}.

(b) Sulphur dioxide as an acidic oxide

(i) *Action of water on sulphur dioxide*

Fill a dry test-tube with sulphur dioxide (from a cylinder) by displacement of air. Invert the tube in distilled water in a small dish and agitate the tube and then leave in an upright position for a few minutes. Would you say that sulphur dioxide is very soluble in water? Test the resulting solution with a few drops of Universal Indicator solution.

(ii) *Action of sodium hydroxide solution on sulphur dioxide*

Repeat Experiment 8.11(b)(i) but use an approximately 2M sodium hydroxide solution in place of the water. Write two equations for the action of OH^- ions on sulphur dioxide.

(c) Sulphur dioxide as a reducing agent

When moist, sulphur dioxide acts as a reducing agent. However, these reactions are best regarded as redox reactions involving the sulphite ion (see Section 8.12).

(d) Sulphur dioxide as an oxidising agent

Strong reducing agents may force sulphur dioxide to act as an oxidising agent. Typical strong reducing agents which act in this manner are hydrogen sulphide (see Section 8.9(d)(i) and magnesium (see p. 120).

Reduction of sulphur dioxide with magnesium

Caution: do not look directly at burning magnesium. Place the boiling-tube in a rack, i.e. do not hold it in the hands.

Wrap a strip of magnesium ribbon (about 4 cm long) round a combustion spoon and ignite it in a Bunsen flame. Immediately lower it cautiously into a boiling-tube that has been previously filled with sulphur dioxide. Examine the contents of the tube after it has cooled down and write an equation for this reaction.

(e) Formation of sulphur trioxide, SO_3

Place about 0·5 g of iron(II) sulphate in a test-tube and heat strongly until no further change takes place. Note the liberation of white fumes (sulphur trioxide). Allow these white fumes to collect in a test-tube containing about 2 cm^3 of distilled water. Shake the contents of this tube and then add about 2 cm^3 of barium chloride solution followed by 2 cm^3 of approximately 2M hydrochloric acid. Note the formation of a white precipitate of barium sulphate—a specific test for sulphate ions (see p. 125).

The reactions taking place in the above experiment are:

$$2Fe^{2+}SO_4^{2-} \rightarrow (Fe^{3+})_2(O^{2-})_3 + SO_3 + SO_2 + 7H_2O$$

$$SO_3 + H_2O \rightarrow H_2SO_4$$

$$Ba^{2+} + SO_4^{2-} \rightarrow Ba^{2+}SO_4^{2-}$$

8.12 Sulphurous Acid and Sulphites

An aqueous solution of sulphurous acid, H_2SO_3, is obtained when sulphur dioxide is passed into water (see Experiment 8.11(b)(i)). The solution contains H^+, HSO_3^- and SO_3^{2-} ions, together with free sulphur dioxide:

$$SO_2 + H_2O \rightleftharpoons H_2SO_3 \rightleftharpoons H^+ + HSO_3^- \rightleftharpoons 2H^+ + SO_3^{2-}$$

Although pure sulphurous acid does not exist, sulphites of the alkali metals, e.g. $(Na^+)_2SO_3^{2-}$, can be obtained as solids. Hydrogen sulphites, e.g. $Na^+HSO_3^-$, also exist in solution, but when attempts are made to isolate them, two hydrogen sulphite ions condense with the elimination of water and a pyrosulphite is deposited:

$$2HSO_3^- \rightleftharpoons S_2O_5^{2-} + H_2O$$

The above reaction is reversible since pyrosulphites give the reactions of hydrogen sulphites in aqueous solution.

(a) Reducing action of sulphurous acid and sulphites

An aqueous solution of sulphurous acid and acidified solutions of sulphites are fairly strong reducing agents as the following standard redox potential indicates:

$$SO_4^{2-} + 2H^+ + 2e^- \rightarrow SO_3^{2-} + H_2O \qquad E^\ominus = +0.20 \text{ V}$$

They should therefore, be capable of reducing chlorine, iodine, iron(III) and acidified potassium permanganate in solution, as the following standard redox potentials indicate:

$$Cl_2 + 2e^- \rightarrow 2Cl^- \qquad E^\ominus = +1.36 \text{ V}$$
$$I_2 + 2e^- \rightarrow 2I^- \qquad E^\ominus = +0.54 \text{ V}$$
$$Fe^{3+} + e^- \rightarrow Fe^{2+} \qquad E^\ominus = +0.76 \text{ V}$$
$$MnO_4^- + 8H^+ + 5e^- \rightarrow Mn^{2+} + 4H_2O \qquad E^\ominus = +1.52 \text{ V}$$

(i) *Reduction of chlorine water*

Dissolve two or three crystals of sodium sulphite in about 3 cm³ of distilled water and add about 2 cm³ of chlorine water. Test for the formation of the sulphate ion (see p. 125) and the chloride ion (see p. 141). Compare these tests with control tests for the sulphate ion (on the original sulphite solution) and for the chloride ion (on the original chlorine water). Is there any difference? Complete and balance the following equation:

$$Cl_2 + SO_3^{2-} + H_2O \rightarrow$$

(ii) *Reduction of an aqueous solution of iodine*

To about 3 cm³ of an aqueous solution of sodium sulphite add about 2 cm³ of a dilute solution of iodine in aqueous potassium iodide. Note the result. Carry out the sulphate test on the resulting solution and compare with the control test on the original sulphite solution. Write an equation for this reaction.

(iii) *Reduction of iron(III) chloride solution*

To about 3 cm³ of a solution of sodium sulphite add about six drops of iron(III) chloride solution and the same number of drops of approximately 2M hydrochloric acid. Note the result and then warm the mixture for about one minute. After cooling, test for the presence of iron(II) ions, Fe^{2+}, by adding approximately 2M sodium hydroxide solution. Complete and balance the following equation:

$$Fe^{3+} + SO_3^{2-} + H_2O \rightarrow$$

(iv) *Reduction of potassium dichromate solution*

To about 2 cm³ of a solution of sodium sulphite add about the same volume of a dilute solution of potassium dichromate, previously acidified with dilute sulphuric acid. Note the result. Complete and balance the equation:

$$Cr_2O_7^{2-} + SO_3^{2-} + H^+ \rightarrow Cr^{3+} + ? + ?$$

(v) *Reduction of potassium permanganate solution*

To about 2 cm³ of a solution of sodium sulphite add about the same volume of a dilute solution of potassium permanganate, previously acidified with dilute sulphuric acid. Note the result. Complete and balance the equation:

$$MnO_4^- + SO_3^{2-} + H^+ \rightarrow Mn^{2+} + ? + ?$$

Tests (iv) and (v) above are frequently used as tests for sulphur dioxide, filter paper strips being dipped into acidified solutions of potassium dichromate or potassium permanganate.

(b) Tests for the sulphite ion

(i) *Action of hydrochloric acid on a solid sulphite*

Sulphur dioxide is liberated (see Experiment 8.11(a)).

(ii) *Action of barium chloride solution on an aqueous solution of a sulphite*

To about 3 cm³ of a solution of sodium sulphite add about the same volume of barium chloride solution and note what happens. Now add approximately 2M hydrochloric acid and note any change. Explain your results.

(c) Action of water on sodium sulphite and sodium pyrosulphite

Dissolve about 0·3 g of sodium sulphite, $(Na^+)_2SO_3^{2-}$, in about 2 cm³ of distilled water and then add 2–3 drops of Universal Indicator solution. Repeat the experiment with sodium pyrosulphite, $(Na^+)_2S_2O_5^{2-}$, which gives the hydrogen sulphite ion, HSO_3^-, in water (see p. 120). Attempt to explain your results by considering the effect of H_2O on the respective ions SO_3^{2-} and HSO_3^-.

8.13 Sulphuric Acid and Sulphates

Essentially the manufacture of sulphuric acid involves the conversion of sulphur dioxide into sulphur trioxide which is then dissolved in water (actually 98% sulphuric acid) to form sulphuric acid (see an inorganic textbook for details of the Contact Process). Sulphur trioxide and sulphuric acid can be formed in a laboratory by heating hydrated iron(II) sulphate (see Experiment 8.11(e)).

Pure sulphuric acid is covalent, its molecule having an approximately tetrahedral structure and containing 6-valent sulphur. It has a high boiling point (543K) and a high viscosity.

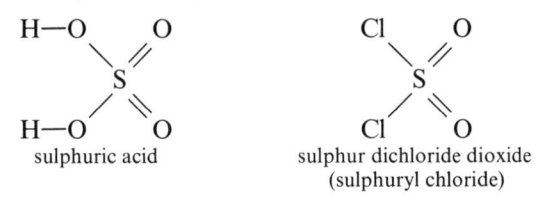

sulphuric acid

sulphur dichloride dioxide (sulphuryl chloride)

Can you account for the high boiling point and high viscosity, c.f. sulphur dichloride dioxide which has a boiling point of 342K and is a mobile liquid but has a higher relative molecular mass than sulphuric acid?

Caution: sulphuric acid is a dangerously corrosive liquid and should be handled with care. If you do spill any accidentally on the skin, wash it off with plenty of water and then neutralise the acid with sodium hydrogen carbonate.

(a) Displacement reactions involving concentrated sulphuric acid

On account of its high boiling point, concentrated sulphuric acid will displace more volatile acids from their salts.

Displacement of hydrogen chloride

Do this reaction in a fume cupboard. Place about 0·3 g of sodium chloride in a test-tube and cautiously add about five drops of concentrated sulphuric acid. Note and explain what happens, writing an equation for the reaction.

(b) Dehydrating properties of concentrated sulphuric acid

Concentrated sulphuric acid has such an affinity for water that it will remove it from compounds with the evolution of much heat. The relevant equations are:

$$H_2SO_4 + H_2O \rightleftharpoons H_3O^+ + HSO_4^- \quad \text{(virtually complete)}$$
$$HSO_4^- + H_2O \rightleftharpoons H_3O^+ + SO_4^{2-} \quad \text{(equilibrium over to left)}$$

(i) *Exothermic reaction between sulphuric acid and water*

Place about 50 cm³ of distilled water in a small beaker and record its temperature. **Cautiously** add about 15 drops of concentrated sulphuric acid, stir well and record the temperature rise.

(ii) *Dehydration of hydrated copper(II) sulphate*

Place about 0·5 g of powdered copper(II) sulphate crystals in a test-tube and cover them with concentrated sulphuric acid. Shake the tube periodically and explain the result.

(iii) *Dehydration of methanoic acid, HCOOH, with sulphuric acid*

Place about 0·5 g of sodium methanoate, $HCOO^-Na^+$, in a test-tube and cover the solid with concentrated sulphuric acid. Warm **very gently** and then apply a lighted taper to the mouth of the test-tube. What gas is evolved? Write an equation for the action of sulphuric acid on sodium methanoate (what type of reaction is this?) and the subsequent reaction between formic acid and sulphuric acid.

(c) Oxidising action of concentrated sulphuric acid

Concentrated sulphuric acid functions as an oxidising agent. A number of reduction products can be formed, their relative proportions in any one particular reaction depending upon such factors as concentration of the acid, strength of the reducing agent present, and temperature. These possible reduction products are shown schematically below (underlined):

$$2H_2SO_4 + 2e^- \rightarrow SO_4^{2-} + 2H_2O + \underline{SO_2}$$

$$5H_2SO_4 + 8e^- \rightarrow 4SO_4^{2-} + 4H_2O + \underline{H_2S}$$

which may be followed by oxidation of the hydrogen sulphide to sulphur:

$$3H_2S + H_2SO_4 \rightarrow 4H_2O + 4\underline{S}$$

(i) Action of concentrated sulphuric acid on potassium bromide

Do this reaction in a fume cupboard. Place about 0·3 g of potassium bromide in a test-tube and add cautiously about five drops of concentrated sulphuric acid. The first product of this reaction is a colourless fuming gas. What is it? What coloured product is eventually formed? Can you detect any sulphur dioxide? Attempt to write an equation for the reaction between the fuming gas and sulphuric acid.

(ii) Action of concentrated sulphuric acid on potassium iodide

Do this reaction in a fume cupboard. Place about 0·3 g of potassium iodide in a test-tube and add cautiously about five drops of concentrated sulphuric acid. What is the fuming colourless gas that is evolved? What is the name of the coloured product? Smell the mouth of the tube **very cautiously**. Can you recognise this smell? Write an equation for the reaction between the fuming gas and sulphuric acid.

Which appears to be the stronger reducing agent, the colourless fuming gas evolved in this reaction or the one evolved in 8.13(c)(i) above?

(d) Sulphuric acid as an acid

When dilute, sulphuric acid shows all the usual properties of acids, i.e. it displaces carbon dioxide from a carbonate, neutralises a base to form a salt and water, and reacts with metals high up in the electrode potential series to form a salt and hydrogen.

Action of sulphuric acid on litmus paper

Place one drop of concentrated sulphuric acid on a piece of blue litmus paper. From the results of this experiment say whether you think the acid contains any H^+ ions. Now add a few drops of distilled water to the original spot of sulphuric acid. Record and explain your observation.

(e) Test for the sulphate ion

Aqueous solutions of sulphates give a white precipitate of barium sulphate on the addition of an aqueous solution of barium chloride. This precipitate is insoluble in dilute hydrochloric acid unlike barium sulphite (see test for sulphite ion p. 122).

Action of barium chloride solution on some sulphates in aqueous solution

To about 3 cm³ of approximately M sulphuric acid add about the same volume of barium chloride solution and note what happens. Now add approximately 2M hydrochloric acid and stir to see if the precipitate dissolves.

Repeat, replacing the dilute sulphuric acid with aqueous solutions of other sulphates, e.g. copper(II) sulphate, sodium sulphate, etc.

$$Ba^{2+} + SO_4^{2-} \rightarrow Ba^{2+}SO_4^{2-}$$

(f) Action of heat on sulphates

Place about 0·3 g of copper(II) sulphate in an ignition tube and heat strongly for several minutes. Record and explain any changes. Is copper(II) sulphate more or less thermally stable than copper(II) nitrate?

Carry out Experiment 8.11(e) with iron(II) sulphate, if you have not previously done it.

8.14 Sulphur Dichloride Oxide (Thionyl Chloride), SOCl₂

Sulphur dichloride oxide is made by reacting sulphur dioxide (or a sulphite) with phosphorus pentachloride, the liquid product then being separated by distillation:

$$SO_2 + PCl_5 \rightarrow SOCl_2 + POCl_3$$

(a) Action of water on sulphur dichloride oxide, SOCl₂

Do this reaction in a fume cupboard with the window pulled well down. Place about 5 cm³ of distilled water in a test-tube and then place the test-tube in a rack. **Very cautiously** add about five drops of sulphur dichloride oxide and note the vigour of the reaction. Hold a piece of filter paper, previously soaked in an acidified solution of potassium dichromate, near the mouth of the test-tube and note the result. Test for the chloride ion (see p. 141) in the resulting solution and then attempt to write an equation for the action of water on sulphur dichloride oxide.

(b) Action of ethanol on sulphur dichloride oxide

Do this reaction in a fume cupboard with the window pulled well down. Place about 2 cm³ of ethanol in a test-tube and then place the test-tube in a rack. **Very cautiously** add about five drops of sulphur dichloride oxide and note the vigour of the reaction. Test for the evolution of sulphur dioxide, as in the previous experiment, with acidified potassium dichromate. Test for the presence of the chloride ion in the resulting solution (see p. 141). Attempt to write an equation for the reaction between ethanol, C_2H_5OH, and sulphur dichloride oxide.

Sulphur dichloride oxide is mainly used in organic chemistry for replacing an —OH group by —Cl. Attempt to write an equation for the action of sulphur dichloride oxide on ethanoic acid, $CH_3.COOH$.

8.15 Chlorosulphonic Acid, $HOSO_2Cl$

Chlorosulphonic acid is obtained by reacting hydrogen chloride with fuming sulphuric acid (sulphuric acid containing some dissolved sulphur trioxide):

$$H_2SO_4 + HCl \rightarrow HOSO_2Cl + H_2O$$

Despite a higher relative molecular mass than sulphuric acid, it has a lower boiling point. Can you explain why?

Action of water on chlorosulphonic acid, $HOSO_2Cl$

Do this reaction in a fume cupboard with the window pulled well down. Place about $5 cm^3$ of distilled water in a test-tube and then place the test-tube in a rack. **Very cautiously** add about five drops of chlorosulphonic acid and note the vigour of the reaction. After the reaction has subsided, divide the solution into two equal portions. Test for the sulphate ion in one portion (see p. 125) and the chloride ion (see p. 141) in the other. Attempt to write an equation for this hydrolysis reaction.

8.16 Sulphur Dichloride Dioxide (Sulphuryl Chloride), SO_2Cl_2

This compound may be obtained by the direct combination of sulphur dioxide and chlorine in the presence of charcoal:

$$SO_2 + Cl_2 \rightarrow SO_2Cl_2$$

Action of water on sulphur dichloride dioxide, SO_2Cl_2

Carry out the reaction exactly as described in Section 8.15 but replace the chlorosulphonic acid by sulphur dichloride dioxide. When the reaction is complete, test for the sulphate and chloride ions in the solution.

How does the action of water on sulphur dichloride dioxide compare with the action of water on chlorosulphonic acid?

8.17 Sodium Thiosulphate, $(Na^+)_2S_2O_3^{2-}.5H_2O$

Thiosulphuric acid cannot be isolated, but its salts are well known; the commonest example being sodium thiosulphate, $(Na^+)_2S_2O_3^{2-}.5H_2O$, which is used in photography for 'fixing' the negative. It is obtained by boiling a solution of sodium sulphite with sulphur, followed by filtration and crystallisation:

$$SO_3^{2-} + S \rightarrow S_2O_3^{2-}$$

(a) Stoichiometric equation for the reaction between aqueous solutions of sodium thiosulphate and iodine

Place 20 drops of a 0·05M solution of iodine (iodine in aqueous potassium iodide) in a test-tube. Using the same clean dropping tube add, drop by drop, a 0·1M solution of sodium thiosulphate, shaking the tube after the addition of each drop. Go on adding the sodium thiosulphate until the colour of the iodine is completely discharged after shaking. Note the number of drops of sodium thiosulphate solution added and then calculate the number of moles of sodium thiosulphate, to the nearest whole number, that react with one mole of iodine. Given that iodine is acting as an oxidising agent and that there are only two products of the reaction, complete and balance the equation:

$$S_2O_3^{2-} + I_2 \rightarrow ? + ?$$

Consult a textbook for the reaction of chlorine and bromine with sodium thiosulphate solution.

(b) Action of hydrochloric acid on sodium thiosulphate

To about 3 cm³ of a solution of sodium thiosulphate add about the same volume of approximately 2M hydrochloric acid. Warm gently and smell the mouth of the tube cautiously, then hold a piece of filter paper previously dipped into an acidified solution of potassium dichromate near the mouth of the tube. What gas is evolved? What happens to the contents of the test-tube? Write an equation for the action of acid on sodium thiosulphate:

$$S_2O_3^{2-} + H^+ \rightarrow$$

(c) Action of sodium thiosulphate solution on a solution of silver nitrate

Place about 3 cm³ of a solution of silver nitrate in a test-tube and then add gradually a solution of sodium thiosulphate until no further change occurs. Note what happens.

The dissolving of the initial precipitate (what is it?) is due to the formation of the complex ion $[Ag(S_2O_3)_2]^{3-}$.

(d) Action of sodium thiosulphate solution on silver iodide

Place about 3 cm³ of a solution of silver nitrate in a test-tube and add about the same volume of a solution of potassium iodide to precipitate the silver iodide. Now add a solution of sodium thiosulphate and note the result. Explain your observations in terms of the equations:

$$AgI \rightleftharpoons Ag^+ + I^-$$
(solid) (in solution)
$$+$$
$$S_2O_3^{2-}$$
$$\updownarrow$$
$$?$$

What bearing does this reaction have on the process of photography?

Look up the structure of the thiosulphate ion, $S_2O_3^{2-}$, which contains linked sulphur atoms.

8.18 Peroxodisulphuric Acid, $H_2S_2O_8$, and Potassium Peroxodisulphate (Potassium Persulphate), $(K^+)_2S_2O_8^{2-}$

The acid can be obtained by reacting 1 mole of hydrogen peroxide with 2 moles of chlorosulphonic acid:

$$HO.SO_2.Cl + H_2O_2 + Cl.SO_2.OH$$
$$\downarrow$$
$$HO.SO_2-O-O-SO_2.OH + 2HCl$$

Like hydrogen peroxide it contains the —O—O— linkage and is a strong oxidising agent. The potassium salt, $(K^+)_2S_2O_8^{2-}$, is better known than the parent acid and it too is a strong oxidising agent. The structure of the persulphate ion, $S_2O_8^{2-}$, is:

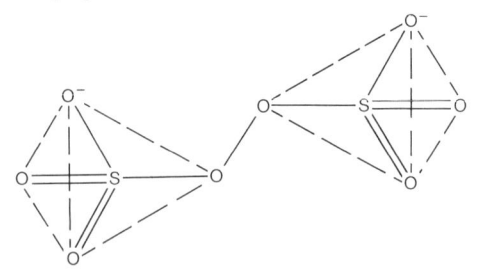

(a) Action of heat on potassium persulphate

Place about 0·3 g of potassium persulphate in an ignition tube and heat it. Test for oxygen with a glowing splint. When no further gas evolution occurs allow the tube to cool and dissolve the residue in a little distilled water. Carry out the test for the sulphate ion on the resulting solution (p. 125). Write an equation for the thermal decomposition of potassium persulphate.

(b) Oxidising action of potassium persulphate

(i) *Oxidation of iron(II) sulphate to iron(III) sulphate*

Place about 3 cm³ of iron(II) sulphate solution in a test-tube and add about the same volume of a solution of potassium persulphate. Warm gently for a few minutes and then cool the solution. Test for the presence of iron(III) ions with approximately 2M sodium hydroxide solution. Write an equation for the reaction between Fe^{2+} and $S_2O_8^{2-}$ ions.

(ii) *Oxidation of potassium iodide solution*

Place about 3 cm³ of a solution of potassium iodide in a test-tube and add about the same volume of a solution of potassium persulphate. Warm gently and note what happens. Write an equation for the action of persulphate ions on iodide ions.

(iii) *Oxidation of copper powder*

Place about 0·2 g of freshly reduced copper powder in a test-tube and then add about 3 cm³ of a solution of potassium persulphate. Now add one drop of silver nitrate solution (to catalyse the reaction) and warm for several minutes. Note the result. Write an equation for the action of persulphate ions on copper powder. Can you suggest how silver ions might catalyse the reaction?

(iv) *Oxidation of manganese(II) sulphate solution*

Place about 2 cm³ of a solution of manganese(II) sulphate in a test-tube and acidify it with about the same volume of approximately M sulphuric acid. Add one drop of silver nitrate solution (as a catalyst) and then about 0·2 g of potassium persulphate. Warm gently and add more potassium persulphate if there is not much evidence of reaction. After the mixture has darkened, filter it and note the colour of the filtrate. What ion do you think is responsible for the colour? Attempt to write an equation for the oxidation of Mn^{2+} by $_2O_8^{2-}$ ions.

8.19 The Halides of Sulphur

The known chlorides of sulphur are disulphur dichloride, S_2Cl_2, sulphur dichloride, SCl_2, and sulphur tetrachloride, SCl_4. Sulphur forms three fluorides with formulae SF_4, SF_6 and S_2F_{10}. Note that sulphur shows valencies of 2 and 4 in the chlorides but 4 and 6 in the fluorides. Look up the boiling points and structures of these halides of sulphur.

Hydrolysis of disulphur dichloride, S_2Cl_2

Add a little disulphur dichloride (prepared in Experiment 8.7(b)) to about 2 cm³ of distilled water and warm gently. The hydrolysis proceeds only slowly, but in time you may be able to detect sulphur dioxide, hydrogen sulphide and hydrochloric acid. Can you detect any sulphur also?

8.20 Summary

(a) Oxygen may be prepared by the decomposition of higher oxides, by the catalytic decomposition of hydrogen peroxide and by the decomposition of many oxysalts. It forms two hydrides, water and hydrogen peroxide.

(b) Water has a polar molecule and will dissolve many ionic salts. Some salt solutions show an acidic reaction, e.g. iron(II) sulphate solution, while others show an alkaline reaction, e.g. sodium carbonate solution. Water containing dissolved ionic compounds may be purified by the process of ion-exchange.

(c) Hydrogen peroxide is unstable with respect to water and oxygen and the addition of manganese(IV) oxide (a catalyst) decomposes an aqueous solution into water and oxygen. An aqueous solution of hydrogen peroxide, acidified with dilute sulphuric acid, will oxidise iron(II) to iron(III) and iodide ions to iodine. The stronger oxidising agent potassium permanganate forces it to assume the rôle of reducing agent.

(d) Sulphur exhibits allotropy, rhombic sulphur being the more stable form below 368·6K and monoclinic sulphur the more stable form above this temperature. Both forms of sulphur contain S_8 molecules, but packed together into different crystal structures.

(e) Sulphur forms a number of hydrides, but only hydrogen sulphide, H_2S, is of any importance. It may be obtained by the action of hydrochloric acid on iron(II) sulphide. Hydrogen sulphide is weakly acidic in aqueous solution. It precipitates many insoluble sulphides from solutions of their salts and is a fairly powerful reducing agent. Aqueous solutions of sulphides of Group 1A metals react with sulphur forming polysulphides, e.g. $(K^+)_2 S_{n+1}^{2-}$.

(f) The only oxides of sulphur of any importance are sulphur dioxide, SO_2, and sulphur trioxide, SO_3. Sulphur dioxide may be prepared by the action of hydrochloric acid on a sulphite. It is a reducing agent in solution (reducing action of the sulphite ion), but both hydrogen sulphide and burning magnesium force it to behave as an oxidising agent. With water it forms sulphurous acid.

(g) Sulphur trioxide may be obtained by the action of heat on iron(II) sulphate. It reacts violently with water forming sulphuric acid.

(h) Sulphurous acid (or moist sulphur dioxide) and acidified solutions of sulphites are strong reducing agents; for example, they convert chlorine to chloride ions and iron(III) ions to iron(II) ions.

(i) Concentrated sulphuric acid will displace more volatile acids from their salts, for example hydrogen chloride from sodium chloride. It will dehydrate many substances, for example methanoic acid to carbon dioxide. In addition, it is a strong oxidising agent; for example, it liberates bromine from hydrogen bromide and iodine from hydrogen iodide, being converted itself into sulphur dioxide (and water) and hydrogen sulphide respectively. Dilute sulphuric acid shows the typical properties of a strong acid.

(j) Sulphites, but not sulphates, will react with hydrochloric acid to give sulphur dioxide.

(k) Sulphur dichloride oxide, $SOCl_2$, is hydrolysed by water to hydrochloric acid and sulphur dioxide. The action of water on chlorosulphonic acid, $HOSO_2Cl$, and sulphur dichloride dioxide, SO_2Cl_2, produces hydrochloric acid and sulphuric acid.

(l) Sodium thiosulphate, $(Na^+)_2S_2O_3^{2-}.5H_2O$, complexes with silver ions and this reaction is used in photography for 'fixing' the negative. It oxidises iodine solutions to iodide ions and can be used for estimating such solutions. Hydrochloric acid reacts with it, forming sulphur dioxide and sulphur.

(m) Peroxodisulphuric acid, $H_2S_2O_8$, is a strong oxidising agent and, like hydrogen peroxide, contains the —O—O— linkage. The potassium salt, $(K^+)_2S_2O_8^{2-}$, is better known than the acid itself and it too is a strong oxidising agent; for example, it oxidises copper to copper(II) ions in solution and also manganese(II) ions to permanganate ions.

(n) Sulphur forms three chlorides with formulae S_2Cl_2, SCl_2 and SCl_4. Disulphur dichloride, S_2Cl_2, is the main product when chlorine is passed over molten sulphur. It is slowly hydrolysed by water forming hydrochloric acid, sulphur, sulphur dioxide and oxyacids of sulphur.

9

GROUP 7B FLUORINE, CHLORINE, BROMINE, IODINE AND ASTATINE

9.1 Some Physical Data of Group 7B Elements

	Atomic number	Electronic configuration	Standard redox potential/V	Density/ g cm^{-3}	M.p./ K	B.p./ K	Atomic radius/nm	Ionic radius/nm X$^-$
F	9	2.7 $1s^2 2s^2 2p^5$	+2·87		50	86	0·72	1·36
Cl	17	2.8.7 $...2s^2 2p^6 3s^2 3p^5$	+1·36		171	239	0·99	1·81
Br	35	2.8.18.7 $...3s^2 3p^6 3d^{10} 4s^2 4p^5$	+1·07	3·12	266	332	1·14	1·95
I	53	2.8.18.18.7 $...4s^2 4p^6 4d^{10} 5s^2 5p^5$	+0·54	4·94	387	456	1·33	2·16
At	85	2.8.18.32.18.7 $...5s^2 5p^6 5d^{10} 6s^2 6p^5$	+0·3					

9.2 Some General Remarks about Group 7B

All members of Group 7B are non-metallic, although there is the usual increase in 'metallic' character with increasing atomic number; for example, dipyridine iodine nitrate can be written as [I(pyridine)$_2$]$^+$NO$_3^-$, containing the I$^+$ ion as part of a complex. Fluorine and chlorine are gases, bromine is a volatile liquid, and iodine is a dark shiny coloured solid. Astatine is radioactive and very short-lived; the small amount of chemistry that has been carried out on this element has employed tracer techniques.

These elements can enter into chemical combination and complete their octets by gaining one electron to form the 1-valent ion, e.g. F$^-$, Cl$^-$ etc. and by forming one covalent bond, e.g. the elements themselves F$_2$, Cl$_2$, Br$_2$, I$_2$, and their hydrides, HF, HCl, HBr and HI.

Because chlorine, bromine and iodine have easily accessible d orbitals available, they are able to form covalent compounds in which the octet of electrons is expanded; for instance, iodine shows valencies of 1, 3, 5 and 7 respectively in the compounds ICl, ICl$_3$, IF$_5$ and IF$_7$. Like nitrogen and oxygen (the first members of Group 5B and 6B respectively) fluorine cannot expand its octet and is thus restricted to a covalency of 1.

The molecules of the halogens are diatomic with only weak van der Waals' forces operating between the individual molecules; however, in the case of iodine these forces are sufficiently strong to bind the iodine molecules into a three dimensional lattice. This structure is easily broken down on heating, and in fact iodine sublimes at one atmosphere pressure if warmed gently.

Chlorine, bromine and iodine form part of a well-graded series and the experimental work in this chapter is restricted to these three elements. Fluorine is exceptionally reactive and this is due to the low bond energy of the fluorine molecule and to the fact that fluorine forms very strong bonds with most other elements; for these reasons fluorine often forces another element in combination with it to display its highest valency, e.g. SF_6. The reason why some metallic fluorides are ionic while the corresponding chlorides are covalent is mainly due to the much greater lattice energy of the ionic fluoride, i.e. the smaller fluoride ion can bind cations more strongly than the larger chloride ion; for example, aluminium fluoride is ionic, but aluminium chloride is covalent.

Other important differences between fluorine and the rest of the halogens are listed below:
  (a) Fluorine oxidises water with the liberation of oxygen, while the other halogens dissolve to give mainly physical solutions (bromine and iodine) or some acid formation in addition (chlorine). No stable oxyacids of fluorine are known.
  (b) Fluorine reacts with alkaline solutions to give oxygen difluoride in the cold and oxygen if the solution is warmed. The other halogens react with alkaline solutions to give an oxysalt, the nature of which depends upon the experimental conditions. No oxysalts containing fluorine are known.
  (c) Hydrogen fluoride is a liquid owing to the presence of hydrogen bonding, whereas the other halogen hydrides are gases. An aqueous solution of hydrogen fluoride is only a weak acid, and this is primarily due to the very high H—F bond energy; the other halogen hydrides function as strong acids in aqueous solution.

9.3 Formation of Chlorine, Bromine and Iodine

The halogens (except fluorine) may be obtained by oxidation of the corresponding hydride with manganese(IV) oxide.

$$MnO_2 + 4HX \rightarrow MnX_2 + X_2 + 2H_2O$$

The experiments should be carried out in a fume cupboard.

(a) Formation of chlorine

Place about 0·3 g of sodium chloride and 0·3 g of manganese(IV) oxide in a test-tube and mix well. Add about 1 cm³ of concentrated sulphuric acid and **warm very gently** and observe the liberation of a light green gas which bleaches moist litmus paper.

(b) Formation of bromine

Repeat the experiment but this time use potassium bromide in place of the sodium chloride. Note the formation of a dark red solution which readily produces a brown vapour.

(c) Formation of iodine

Repeat the experiment but replace the potassium bromide with potassium iodide.

9.4 Solubility of the Halogens in Water and in Organic Solvents

Chlorine and bromine are moderately soluble in water (bromine more so than chlorine), while iodine is only very sparingly soluble. Although hydrolysis of the halogens takes place in solution, this is only extensive in the case of chlorine, i.e. aqueous solutions of bromine and iodine are essentially solutions containing respectively Br_2 and I_2 molecules.

It is possible to discuss the dissolution of chlorine in water in terms of two equilibrium reactions:

$$Cl_2(g) + Water \rightleftharpoons Cl_2(\text{in solution})$$

$$Cl_2(\text{in solution}) + 2H_2O \rightleftharpoons H_3O^+ + Cl^- + HOCl$$
$$\text{chloric(I) acid}$$

The dissociation constant for the second reaction, K, is given by the usual expression:

$$K = \frac{[H_3O^+][Cl^-][HOCl]}{[Cl_2(\text{in solution})]} = 4.2 \times 10^{-4} \text{ mol}^2 \text{dm}^{-6} \text{ (at 298K)}$$

For bromine and iodine the values of K are respectively 7.2×10^{-9} and 2.0×10^{-13}, thus bromic(I) acid, HOBr, and iodic(I) acid, HOI, are formed in negligible amounts.

Although iodine is only sparingly soluble in water, it dissolves appreciably in potassium iodide solution, since reaction occurs with the formation of the complex I_3^- ion:

$$I_2 + I^- \rightleftharpoons I_3^-$$

(a) Solubility of iodine in potassium iodide solution

Place a few small crystals of iodine in a test-tube and then add about 2 cm³ of distilled water. Warm gently and then allow to cool. Note that very little iodine dissolves. Now add about 2 cm³ of potassium iodide solution and note that much more iodine dissolves with the formation of a dark brown solution.

(b) Solubility of the halogens in tetrachloromethane

Place about 3 cm³ of freshly prepared chlorine water in a test-tube and then add about 0·5 cm³ of tetrachloromethane. Cork the tube and shake. Observe the lower tetrachloromethane layer and explain the result.

Repeat the experiment with bromine water but this time use about 2 cm³ of the bromine solution. Observe and explain what happens.

Repeat with a solution of iodine in potassium iodide solution, noting and explaining your observations.

9.5 Reactions of the Halogens with some Elements

Chlorine, bromine and iodine combine with many metals and non-metals; in general, the combination with chlorine is the most vigorous and with iodine the least vigorous, although there are some notable exceptions to this rule. For example potassium explodes with bromine and iodine. Exceptions such as these can be attributed to a greater halogen concentration in liquid bromine and solid iodine as compared with gaseous chlorine. Carbon, nitrogen and oxygen do not combine with any of the halogens directly.

(a) Reactions of the halogens with some metals

(i) *The combination of chlorine with magnesium, aluminium and tin*

These reactions have been described previously. See Experiments 3.4(b), 4.8(a) and 5.15(a)(i).

(ii) *The combination of bromine with aluminium and tin*

These experiments should be carried out in a fume cupboard and care should be exercised when handling liquid bromine since it is a dangerous and corrosive substance.

Place about 0·5 cm³ of liquid bromine into each of two test-tubes using a dropping tube and then stand the test-tubes in a test-tube rack. Heat a piece of aluminium foil (about 1 cm × 3 cm) and then drop it into one of the test-tubes. Note the vigour with which it combines with bromine.

Repeat the experiment with a similar sized piece of tin foil, but this time drop it into the bromine without heating it. It should combine spontaneously. The product is tin(IV) bromide, $SnBr_4$.

(iii) *The combination of iodine with aluminium*

This experiment has been described previously. See Experiment 4.5(b), p. 42.

(b) Reactions of the halogens with some non-metals

(i) *The combination of chlorine with phosphorus and sulphur*

The reaction of chlorine with sulphur has been described previously.

See Experiment 8.7(b), p. 113.

The combination of chlorine with phosphorus should be done as **a demonstration in a fume cupboard.** Place a small piece of white phosphorus in a deflagrating spoon and then lower it into a gas jar containing dry chlorine. Note that the phosphorus inflames with the formation of a white solid (phosphorus pentachloride, PCl_5).

(ii) *The combination of bromine with red phosphorus*

Place about $0.5\ cm^3$ of liquid bromine in a test-tube and then stand the test-tube in a test-tube rack in a fume cupboard. Add a very small portion of **red phosphorus (it is dangerous to add white phosphorus)** and note the spontaneous reaction. If bromine is in excess, the product is phosphorus pentabromide, PBr_5.

9.6 The Action of Alkalis and Acids on Aqueous Solutions of the Halogens

Chlorine, bromine and iodine react to some extent with water giving rise to the following general equilibrium:

$$X_2(\text{in solution}) + 2H_2O \rightleftharpoons H_3O^+ + X^- + HOX$$

where X stands for the halogen atom. However, as pointed out in Section 9.4, this reaction is not very extensive for both bromine and iodine.

(a) Action of sodium hydroxide solution and dilute sulphuric acid on the above equilibrium reaction

Place about $3\ cm^3$ of freshly prepared chlorine water in a test-tube and note the faint colour of the solution. Now add, drop by drop, an approximately 2M solution of sodium hydroxide. Does the colour fade? Now add a few drops of approximately M sulphuric acid and note if the faint colour returns when the acid is in excess.

Repeat the experiment with bromine water and add the sodium hydroxide solution until the colour of the solution is almost discharged. Now add approximately M sulphuric acid until the acid is in excess.

Repeat again, but this time use a solution of iodine in potassium iodide solution.

Attempt to explain these colour changes by discussing the effect of alkali and acid on the above general equilibrium reaction.

(b) Action of chlorine on concentrated potassium hydroxide solution (Demonstration)

Dissolve some pellets of potassium hydroxide in about $10\ cm^3$ of distilled water and place the concentrated solution in a boiling-tube. Pass chlorine into the solution (**in a fume cupboard**) and note that the solution goes yellow as potassium chlorate(I), K^+ClO^-, is being formed. Go on passing chlorine into the solution and note that the yellow colour disappears and that the temperature rises as the chlorate(I) is converted into potassium chlorate(V), $K^+ClO_3^-$. Cool the mixture and note the precipitation of potassium chlorate(V) crystals.

The reactions can be explained as follows:
At first the chlorine reacts with hydroxyl ions to give the chlorate(I):

$$2OH^- + Cl_2 \rightarrow Cl^- + ClO^- + H_2O$$

As the temperature rises there is a tendency for the chlorate(I) ion to disproportionate into the chloride and the chlorate(V) thus:

$$3OCl^- \rightarrow 2Cl^- + ClO_3^-$$

The overall reaction is:

$$6OH^- + 3Cl_2 \rightarrow 5Cl^- + ClO_3^- + 3H_2O$$

Bromine and iodine react with hydroxyl ions in a similar manner but there is an increasing tendency for the bromate(I) and iodate(I) ions to disproportionate into the bromate(V) and iodate(V) ions respectively:

$$3OBr^- \rightarrow 2Br^- + BrO_3^-$$
$$3OI^- \rightarrow 2I^- + IO_3^-$$

9.7 Oxidising Reactions of the Halogens

Chlorine is a powerful oxidising agent and will remove hydrogen from hydrocarbons. Bromine reacts in a similar manner, but iodine shows little reaction under the same conditions.

(a) Reaction of chlorine with ethyne (acetylene) (Demonstration)

Carry out this experiment in a fume cupboard. Place about 1 g of bleaching powder in a boiling-tube and add two or three small lumps of calcium carbide. Now add, from a dropping tube, a small amount of concentrated hydrochloric acid. Ethyne and chlorine are evolved and react with the production of a flame and a black smoke of carbon:

$$H-C\equiv C-H + Cl_2 \rightarrow 2C + 2HCl$$

(b) Reaction of bromine with benzene

Place about five drops of benzene in a test-tube and then add some iron filings. Now add a few drops of bromine and leave for a few minutes. Heat should be generated and the evolution of the fuming gas hydrogen bromide should occur:

$$C_6H_6 + Br_2 \rightarrow C_6H_5Br + HBr$$

(c) Displacement reactions with the halogens

Place about 2 cm³ of a solution of sodium chloride in a test-tube. Into a second tube place an equal volume of potassium bromide solution and into a third place the same amount of potassium iodide solution. To each tube add $2\,cm^3$ of a freshly prepared solution of chlorine water and note any changes. Now add about $1\,cm^3$ of tetrachloromethane to each tube, then cork and shake. Note any changes in the lower organic layer and explain them.

Do your results agree with the behaviour predicted from the values of the standard redox potentials of chlorine, bromine and iodine given in Section 9.1, p. 132? Write an ionic equation for the reaction of chlorine with iodide ions. How would you expect bromine water to react with aqueous solutions of sodium chloride and potassium iodide? Carry out these two experiments to check your answers.

(d) Other oxidising reactions of the halogens

(i) *Reaction with hydrogen sulphide*

Do these experiments in a fume cupboard. Pass a stream of hydrogen sulphide through about $2\,cm^3$ of aqueous solutions of chlorine and bromine contained in separate test-tubes. Repeat with a solution of iodine in potassium iodide solution. Note in each case the formation of a precipitate. What is it? What have the halogens been converted into?

(ii) *Reaction with sulphites*

To about $2\,cm^3$ of aqueous solutions of chlorine and bromine contained in separate test-tubes add a solution of sodium sulphite, until the colour of the halogen has been discharged. Repeat with a solution of iodine in potassium iodide solution, again adding the sulphite solution until the colour of the halogen has been discharged. Now test for the presence of the sulphate ion in each solution (see Experiment 8.13(e), p. 125). Also do a control test using an equal volume of the sulphite solution since there may well be some initial sulphate present. Complete and balance the general equation:

$$X_2 + SO_3^{2-} + H_2O \rightarrow$$

9.8 The Halogen Hydrides

Chlorine, bromine and iodine combine under appropriate conditions with hydrogen to give the halogen hydride:

$$H_2 + X_2 \rightarrow 2HX$$

Chlorine explodes violently with hydrogen when the mixture is irradiated with ultraviolet light, but a stream of hydrogen can be burnt safely in chlorine.

Hydrogen and bromine combine smoothly at about 600K in the presence of a platinum catalyst to give hydrogen bromide, HBr, but under the same conditions the reaction between hydrogen and iodine is reversible.

(a) Combination of hydrogen and chlorine

(i) *Burning hydrogen in chlorine* **(Demonstration)**

Allow hydrogen from a cylinder to pass through a glass tube that has been bent into a semicircle and fitted with a fireclay jet. After allowing the gas to flow for several seconds, ignite it, then lower the burning jet into a gas jar of chlorine. The flame should turn green producing hydrogen chloride which fumes in moist air.

(ii) *Explosion of a chlorine/hydrogen mixture* **(Demonstration)**

Fill a small polythene bottle (the type used for indicator solutions) with an equal volume of hydrogen and chlorine (over brine) and then insert a rubber bung. Stand the bottle on the bench and shine a photofloodlight onto it. There should be a small explosion which propels the bung out of the polythene bottle.

(b) Formation of the halogen hydrides by the action of concentrated sulphuric acid on halides

(i) *Action of concentrated sulphuric acid on sodium chloride*

Place about 0·1 g of sodium chloride in a test-tube and then add about 0·5 cm^3 of concentrated sulphuric acid. Warm **very gently** and note the liberation of a fuming gas which is hydrogen chloride.

(ii) *Action of concentrated sulphuric acid on potassium bromide*

Repeat the above experiment, but use about 0·1 g of potassium bromide in place of the sodium chloride. What else do you observe in addition to the liberation of the fuming gas hydrogen bromide? Hold a piece of filter paper, previously dipped into an acidified solution of potassium dichromate, near the mouth of the tube and also a piece of filter paper previously dipped into a solution of lead(II) ethanoate. What happens and why? Attempt to write an equation for any reaction between the hydrogen bromide which is liberated and the concentrated sulphuric acid.

(iii) *Action of concentrated sulphuric acid on potassium iodide*

Repeat the above experiment, but use 0·1 g of potassium iodide in place of the potassium bromide. What do you observe in addition to the liberation of the fuming gas hydrogen iodide? Hold a piece of filter paper, previously dipped into a solution of lead(II) ethanoate, near the mouth of the tube. What happens and why? Attempt to write an equation for any reaction between the liberated hydrogen iodide and the concentrated sulphuric acid.

From your observations in the above three experiments can you arrange the halogen hydrides in increasing order as reducing agents?

(c) Formation of the halogen hydrides by the action of phosphoric acid on halides, and some of their reactions

Place about 2 g of sodium chloride in a boiling-tube and add about 2 cm³ of orthophosphoric acid followed by about 1 g of phosphorus(V) oxide (to absorb any water in the acid). Place a rubber bung containing glass tubing, bent through two right angles, into the boiling-tube as shown in Fig. 9.1 and warm the mixture gently. Collect the evolved hydrogen chloride in three dry test-tubes, sealing the tubes with corks when they appear to be full, i.e. when fumes appear at the top of the test-tubes.

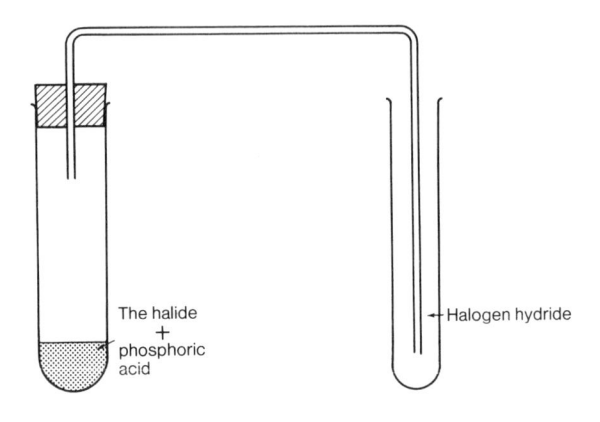

Fig. 9.1
Formation of the halogen hydrides.

Carry out the following experiments on the hydrogen chloride.

(i) *Solubility of hydrogen chloride in water*

Open one of the test-tubes under a beaker of water and note whether the water level rises rapidly. What can you deduce from this experiment? Add a few drops of Universal Indicator solution to the solution in the test-tube, having removed it from the beaker with the thumb over the open end. What happens and why?

(ii) *Thermal stability of hydrogen chloride*

Heat a nichrome wire until it is red hot and then place it into a test-tube containing hydrogen chloride. Does it have any apparent effect? If not, try a hotter piece of wire.

(iii) *Reaction with ammonia*

Dip a glass rod into concentrated ammonia solution and hold it in a test-tube containing hydrogen chloride. What happens and why?

Prepare some hydrogen bromide using the apparatus shown in Fig. 9.1, replacing the 2 g of sodium chloride by the same mass of potassium bromide. Collect the gas evolved in three test-tubes. How does the action of phosphoric acid on potassium bromide differ from the action of sulphuric acid on this salt? Can you explain why?

Carry out Experiments (i), (ii) and (iii) above on the hydrogen bromide.

Prepare some hydrogen iodide in a similar manner using 2 g of potassium iodide in place of the potassium bromide. Again carry out Experiments (i), (ii) and (iii).

From the results of Experiment (ii) carried out on all three gases can you arrange these halogen hydrides in increasing order of thermal stability?

9.9 Tests for the Halide Ions

(a) Displacement from solution

Chlorine will displace bromine from an aqueous solution of a bromide, and iodine from an aqueous solution of an iodide (see Experiment 9.7(c), p. 137).

(b) Action of concentrated sulphuric acid on the halide

Concentrated sulphuric acid liberates hydrogen chloride from a chloride, hydrogen bromide (with some bromine and sulphur dioxide) from a bromide and hydrogen iodide (with some iodine and hydrogen sulphide) from an iodide (see Experiments 9.8(b)(i), (ii), (iii), p. 139).

(c) Action of silver nitrate solution on a solution of a halide

To about 1 cm^3 of a dilute solution of sodium chloride add about the same volume of dilute silver nitrate solution and note the colour of the precipitate. Examine the effect of aqueous ammonia on the precipitate.

Repeat the experiment with a dilute solution of potassium bromide. Is the precipitate of silver bromide soluble in bench aqueous ammonia? If not, try a more concentrated ammonia solution.

Repeat again, but this time use a dilute solution of potassium iodide. Can you distinguish between chlorides, bromides and iodides (a) by the colour of their silver salts, (b) by their solubilities in aqueous ammonia?

9.10 The Oxides of Chlorine, Bromine and Iodine

The known oxides of chlorine are dichlorine oxide, Cl_2O (an orange coloured gas), chlorine dioxide ClO_2 (a yellow gas), dichlorine hexoxide, Cl_2O_6 (a red liquid), and dichlorine heptoxide, Cl_2O_7 (a colourless liquid). All four oxides are unstable and dangerously explosive.

The oxides of bromine have not been studied very extensively since they readily decompose into bromine and oxygen at low temperatures, e.g. at 233 K or even lower. Oxides with formulae Br_2O, Br_3O_8, BrO_2 and Br_2O_7 have been reported.

The best known oxide of iodine is iodine(V) oxide, I_2O_5, which is thermally stable up to a temperature of about 573 K.

(a) Some reactions of iodine(V) oxide, I_2O_5

(i) *Action of heat on iodine(V) oxide*

Place about 0·1 g of iodine(V) oxide in an ignition tube and warm gently. Test for oxygen with a glowing splint. Do not heat the tube for long, since the vapour is very irritating.

(ii) *Action of iodine (V) oxide on hydrogen sulphide*

Place about 0·1 g of iodine(V) oxide in a test-tube and then pass hydrogen sulphide into it for a few seconds. What do you observe? Attempt to write an equation for this reaction.

(iii) *Action of water on iodine(V) oxide*

Place about 0·1 g of iodine(V) oxide in a test-tube and then add about 2 cm³ of distilled water. Shake and then add two or three drops of Universal Indicator solution. Note what you observe.

9.11 The Action of Water on some Chlorides

Some chlorides give a neutral solution in water while others give an acidic reaction.

Try the effect of water on the chlorides of sodium, magnesium, aluminium, carbon, silicon and phosphorus.

(a) Action of water on sodium chloride and magnesium chloride

Place about 0·2 g of each chloride in separate test-tubes and then add about 2 cm³ of distilled water to each. Shake and then add two or three drops of Universal Indicator solution. What type of reaction do these chlorides have in water?

(b) Action of water on aluminium chloride, tetrachloromethane, silicon tetrachloride and phosphorus pentachloride

Carry out the experiments described in Sections 4.8(b)(ii), 5.9(a) and (b), and 6.12(a), if you have not done so previously. What type of reaction do these chlorides show with water?

9.12 The Oxyacids of the Halogens and their Salts

(a) The halic(I) acids, HOX, and the halates(I), OX⁻

The halic(I) acids have already been mentioned in Section 9.4. The only halate(I) of any importance is sodium chlorate(I). It is sold under a variety of trade names like Parazone, Domestos and Milton as a dilute aqueous solution (which also contains sodium chloride) for use as a mild disinfectant and bleaching agent.

Sodium chlorate(I) as an oxidising agent in acid solution

Place about 1 cm³ of sodium chlorate(I) solution in a test-tube and acidify it with a few drops of approximately M sulphuric acid. Add a solution of potassium iodide, drop by drop, and observe what happens. Complete and balance the equation:

$$I^- + H^+ + OCl^- \rightarrow$$

Place about 1 cm³ of sodium chlorate(I) in a test-tube and acidify it with a few drops of approximately M sulphuric acid. Add about 0.5 cm³ of a solution of iron(II) sulphate and warm for a few seconds. Cool and then test for the presence of iron(III) by the addition of approximately 2M sodium hydroxide solution. Complete and balance the equation:

$$Fe^{2+} + H^+ + OCl^- \rightarrow$$

(b) The halic(V) acids, HXO₃, and the halates(V), XO₃⁻

The halic(V) acids dissociate extensively in aqueous solution, i.e. they are strong acids. They also behave as strong oxidising agents. Like the parent acids, the halates(V) are oxidising agents.

(i) *The action of heat on the potassium halates(V)*, $K^+XO_3^-$

Place about 0.1 g of potassium chlorate(V) in an ignition tube and heat it. Test for oxygen with a glowing splint, but **do not allow the splint to drop into the ignition tube.** Allow the tube to cool and then add a little distilled water. Carry out the test for the chloride ion on this residue (see Experiment 9.9(c), p. 141).

Repeat the experiment using potassium bromate(V) and potassium iodate(V) in separate ignition tubes.

Write a general equation for these thermal decomposition reactions.

(ii) *Action of acidified solutions of halates(V) on a solution of iron(II) sulphate*

Acidify about 2 cm³ of a solution of potassium chlorate(V) with approximately M sulphuric acid and then add about 1 cm³ of a solution of iron(II) sulphate. Warm for a few seconds, cool and then test for the presence of iron(III) ions with an approximately 2M solution of sodium hydroxide.

Repeat, but replace the potassium chlorate(V) solution with the same volume of potassium bromate(V) solution.

Repeat again, but this time use a solution of potassium iodate(V).

Attempt to write a general equation for the action of an acidified solution of an halate(V) on a solution of iron(II) ions (you will find it simpler to construct this equation using partial equations first).

(iii) *Action of acidified solutions of halates(V) on hydrogen sulphide*

Acidify about 2 cm^3 of a solution of potassium chlorate(V) with approximately M sulphuric acid and then pass hydrogen sulphide through it for several seconds. Can you detect any precipitate? If so what do you think it is?

Repeat, but replace the potassium chlorate(V) solution with the same volume of potassium bromate(V) solution. What causes the solution to turn dark?

Repeat again, but this time use a solution of potassium iodate(V).

Attempt to write a general equation for the action of hydrogen sulphide on an acidified solution of an halate(V).

(iv) *Action of acidified solutions of halates(V) on halides*

Place about 2 cm^3 of a solution of potassium bromate(V) in a test-tube and add about the same volume of a solution of potassium bromide. Acidify the solution with about the same volume of approximately M sulphuric acid and note the result.

Repeat the experiment, but replace the solutions of potassium bromate(V) and potassium bromide by solutions of potassium iodate(V) and potassium iodide. Note what happens.

Attempt to balance the equation:

$$IO_3^- + I^- + H^+ \rightarrow$$

You will be able to check it when you have carried out the experiment in the next section. The reaction between bromate(V) and bromide ions in dilute acid is similar. There is no significant reaction between chlorate and chloride ions in acid solution.

9.13 Potassium Iodate(V), $K^+IO_3^-$, as a Primary Standard

A volumetric reagent which can be made up to an accurately known concentration in water by direct weighing is known as a primary standard. Such a substance must clearly satisfy the following conditions: be obtainable in a high degree of purity, be stable at normal temperatures (room temperature to 373K), and be reasonably soluble in water. Potassium iodate(V) is such a substance and may be used to standardise a solution of sodium thiosulphate. However, in the following experiment, it is used to establish the stoichiometric equation for the reaction between iodate(V) ions and iodide ions in dilute acid.

Stoichiometric equation for the reaction between iodate(V) ions and iodide ions in acidic solution

Place 30 drops of 0·0167M potassium iodate(V) solution in a test-tube, add about 0·2 g of solid potassium iodide (this represents a large excess) and then acidify the resulting solution with 10 drops of approximately M sulphuric acid. Using the same clean dropping tube add, drop by drop, a 0·1M solution of sodium thiosulphate, until the colour of iodine is completely discharged after shaking. Note the number of drops of thiosulphate solution needed (you should be able to obtain this amount to the nearest drop). Calculate, to the nearest whole number, the number of moles of sodium thiosulphate that react indirectly with one mole of potassium iodate(V).

Given that iodine and thiosulphate react as follows:

$$I_2 + 2S_2O_3^{2-} \rightarrow 2I^- + S_4O_6^{2-}$$

calculate the number of moles of iodine that are produced when 1 mole of iodate(V) reacts with excess iodide in the presence of hydrogen ions. Hence construct the equation for the action of iodate(V) ions on iodide ions in the presence of hydrogen ions.

9.14 The Extraction of Iodine

The main source of iodine is sodium iodate(V), $Na^+IO_3^-$, which occurs in the sodium nitrate deposits in Chile. After crystallisation of the sodium nitrate from solution, the mother liquor is treated with a solution of sodium hydrogen sulphite when reduction of iodate(V) to iodine occurs.

In the following experiment either sodium iodate(V) or potassium iodate(V) may be used.

Reaction between solutions of potassium iodate(V) and sodium hydrogen sulphite

Place about 2 cm^3 of a solution of potassium iodate(V) in a test-tube and add gradually a solution of sodium hydrogen sulphite (obtained by dissolving a little sodium pyrosulphite in water, see p. 122). Note what you observe.

Repeat the experiment in reverse, i.e. to about 2 cm^3 of a solution of sodium hydrogen sulphite add a solution of potassium iodate(V) gradually, until no further change occurs.

Can you account for any differences between the two experiments?

9.15 The Interhalogen Compounds

The known interhalogen compounds are listed overleaf; they are covalent substances and polarised in such a way that a fractional negative charge resides on the lighter halogen atom(s). All the compounds listed can be obtained by direct synthesis under the right conditions.

ClF (gas)	ClF$_3$ (gas)		
BrF (gas) BrCl (gas)	BrF$_3$ (liquid)	BrF$_5$ (liquid)	
ICl (liquid) IBr (solid)	ICl$_3$ (solid)	IF$_5$ (liquid)	IF$_7$ (gas)

The interhalogen compounds are oxidising agents; the halogen fluorides, except iodine heptafluoride which is a mild fluorinating agent, are amongst the most reactive substances known; for example bromine trifluoride will react vigorously even with asbestos.

Formation of iodine monochloride, ICl, and iodine trichloride, ICl$_3$ (Demonstration)

This experiment should be carried out in a fume cupboard. Place a few crystals of iodine in a test-tube and direct a steady stream of chlorine (preferably from a cylinder) on to the iodine. The first product is iodine monochloride; what colour is it? Go on passing the chlorine on to the iodine monochloride and note the formation of another product, iodine trichloride, ICl$_3$. What colour is this product? When all the iodine has been converted into iodine trichloride disconnect the chlorine supply and tilt the test-tube slightly, so that the open end is at a lower level than the closed end. What happens? Now pass in more chlorine and note the result. Can you explain these last two observations in terms of a reversible reaction?

9.16 Summary

No experimental work on fluorine and its compounds is included in this book and the following summary refers to the chemistry of chlorine, bromine and iodine only.

(a) The halogens may be prepared by oxidising the relevant halogen hydride with manganese(IV) oxide.

(b) Chlorine and bromine are moderately soluble in water, while iodine is only very sparingly soluble. Iodine is quite soluble in potassium iodide solution with which it forms the I_3^- complex ion. All three halogens are appreciably soluble in organic solvents such as tetrachloromethane.

(c) The halogens react with a variety of metals and non-metals. In general, the reactivity of the halogen decreases in the order chlorine, bromine, iodine.

(d) The halogens react with strong alkaline solutions to give the halate(I) ion, XO^-, and the halide ion, X^-. The tendency for the halate(I) ion, XO^-, to disproportionate into the halate(V) ion, XO_3^-, and the

halide ion, X^-, increases with increasing size of the halogen atom.
(e) The halogens are good oxidising agents; for example, in solution they oxidise hydrogen sulphide to sulphur and aqueous solutions of sulphites to sulphates, themselves being converted into halide ions. Their oxidising ability decreases in the order chlorine, bromine, iodine.
(f) The halogen hydrides may be prepared by heating the appropriate solid halide with orthophosphoric acid. They are all strongly fuming gases and are very soluble in water forming strongly acidic solutions. Their thermal instability and reducing properties increase in the order hydrogen chloride, hydrogen bromide, hydrogen iodide.
(g) The halide ions can be detected in solution by the formation of precipitates with silver nitrate solution. Silver chloride is white, silver bromide is cream and silver iodide is yellow. The solubilities of the silver halides decrease in the order silver chloride, silver bromide, silver iodide.
(h) Iodine(V) oxide, I_2O_5, is the most thermally stable of the halogen oxides; however, on heating to about 573K it decomposes into iodine and oxygen. It oxidises hydrogen sulphide to sulphur and reacts with water, forming a strongly acidic solution containing iodic(V) acid, HIO_3.
(i) The only halate(I) of any importance is sodium chlorate(I), Na^+OCl^-. It oxidises solutions of iodide ions to iodine and iron(II) ions to iron(III) ions.
(j) The halic(V) acids, HXO_3, are strong oxidising agents and in solution they give rise to strongly acidic solutions. The halates(V) are also oxidising agents, e.g. the action of heat on potassium halate(V) gives the halide and oxygen, and acidified solutions of halates(V) convert iron(II) ions to iron(III) ions. Potassium iodate(V) is a primary standard in volumetric analysis, and an acidified solution reacts with potassium iodide solution producing a quantitative yield of iodine. The liberated iodine may be titrated with sodium thiosulphate solution, and hence the strength of this solution may be found.
(k) There are numerous interhalogen compounds which contain polarised covalent bonds. Iodine monochloride, ICl, and iodine trichloride, ICl_3, may be obtained by the action of chlorine on solid iodine. Iodine trichloride exists as an equilibrium mixture with iodine monochloride and chlorine.

10

THE FIRST TRANSITION SERIES (SCANDIUM, TITANIUM, VANADIUM, CHROMIUM, MANGANESE, IRON, COBALT AND NICKEL)

10.1 Some Physical Data of First Transition Series Elements

	Atomic number	Electronic configuration	Standard redox potential/V $M^{2+}+2e^-\rightarrow M$	Standard redox potential/V $M^{3+}+3e^-\rightarrow M$	Density/ $g\,cm^{-3}$	M.p./ K	B.p. K	Atomic radius/ nm	Ionic radius/ nm M^{3+}
Sc*	21	2.8.8(1).2 $...3s^23p^63d^14s^2$		-2.10	3.10			0.144	0.081
Ti	22	2.8.8(2).2 $...3s^23p^63d^24s^2$	-1.63	-1.21	4.43	1953	3808	0.132	0.076
V	23	2.8.8(3).2 $...3s^23p^63d^34s^2$	-1.18	-0.85	6.07	1983		0.122	0.074
Cr	24	2.8.8(5).1 $...3s^23p^63d^54s^1$	-0.91	-0.74	7.19	2163	2755	0.117	0.069
Mn	25	2.8.8(5).2 $...3s^23p^63d^54s^2$	-1.18	-0.28	7.21	1520	2310	0.117	0.066
Fe	26	2.8.8(6).2 $...3s^23p^63d^64s^2$	-0.44	-0.04	7.87	1801	3008	0.116	0.064
Co	27	2.8.8(7).2 $...3s^23p^63d^74s^2$	-0.28	$+0.40$	8.70	1763	3153	0.116	0.063
Ni	28	2.8.8(8).2 $...3s^23p^63d^84s^2$	-0.25		8.90	1725	3163	0.115	0.062
Cu*	29	2.8.8(10).1 $...3s^23p^63d^{10}4s^1$	$+0.34$		8.92	1356	2868	0.117	

*The element copper is not a transition element but Cu^{2+} has transitional characteristics. Sc^{3+} is also non-transitional. See Section 10.2 for a definition of a transition element.

10.2 Some General Remarks about First Transition Series Elements

A transition element may be defined as one that has a partially filled d shell, i.e. the atom of the element has between one and nine d electrons; this definition is extended to include the oxidation (valency) states of these elements which also have partially filled d shells. The first transition series runs from scandium to copper; the element zinc, which immediately follows copper in the Periodic Table, has a full complement of d electrons and so has the zinc ion, so that this element is non-transitional. The outer electronic configurations of the atoms and some ions of the first transition series are given in Table 10A.

TABLE 10A. The outer electronic configurations of the atoms and some ions of the first transition series

Sc	$3p^6 3d^1 4s^2$	Sc^{3+}	$3p^6$(NT)		
Ti	$3p^6 3d^2 4s^2$	Ti^{2+}	$3p^6 3d^2$	Ti^{3+}	$3p^6 3d^1$
V	$3p^6 3d^3 4s^2$	V^{2+}	$3p^6 3d^3$	V^{3+}	$3p^6 3d^2$
Cr	$3p^6 3d^5 4s^1$	Cr^{2+}	$3p^6 3d^4$	Cr^{3+}	$3p^6 3d^3$
Mn	$3p^6 3d^5 4s^2$	Mn^{2+}	$3p^6 3d^5$	Mn^{3+}	$3p^6 3d^4$
Fe	$3p^6 3d^6 4s^2$	Fe^{2+}	$3p^6 3d^6$	Fe^{3+}	$3p^6 3d^5$
Co	$3p^6 3d^7 4s^2$	Co^{2+}	$3p^6 3d^7$	Co^{3+}	$3p^6 3d^6$
Ni	$3p^6 3d^8 4s^2$	Ni^{2+}	$3p^6 3d^8$		
Cu	$3p^6 3d^{10} 4s^1$(NT)	Cu^{2+}	$3p^6 3d^9$	Cu^+	$3p^6 3d^{10}$(NT)
Zn	$3p^6 3d^{10} 4s^2$(NT)			Zn^{2+}	$3p^6 3d^{10}$(NT)

(NT)—non-transitional

There are five separate $3d$ orbitals, each of which can accommodate two electrons with opposite spins. These five energy levels are degenerate, i.e. have the same energy, in the free atoms and ions. In practice each of these separate energy levels must be singly occupied before electron pairing takes place (Hund's rule); for example, in the manganese(II) ion, Mn^{2+}, each of the five $3d$ levels are singly occupied while in the Fe^{2+} ion four of the five $3d$ levels are singly occupied and the fifth holds two electrons with opposite spins.

When all five $3d$ levels are either singly or doubly filled, a degree of stability is conferred on the atom or ion, and this explains why the electronic configurations of the atoms of chromium and copper are respectively $3d^5 4s^1$ and $3d^{10} 4s^1$. It also explains why Fe^{2+} ($3d^6$) is easily oxidised to Fe^{3+} ($3d^5$) and why Mn^{2+} ($3d^5$) is not easily oxidised to Mn^{3+} ($3d^4$).

10.3 The Concept of Oxidation Numbers

The concept of oxidation numbers (and oxidation states) is widely used in transition metal chemistry, but it is also applicable to the chemistry of other elements. Consider manganese(VII) oxide, Mn_2O_7, which is an explosive covalent compound. If electron transfer took place completely from the less to the more electronegative atom, i.e. from manganese to oxygen, the manganese atom would carry a charge of $+7$ units, since two electrons would be transferred to each oxygen atom (a total of fourteen altogether). In this example the oxidation number of manganese is $+7$, and is cometimes written Mn(VII). The oxidation number of an element in a compound is simply obtained by considering how many electrons would be gained or lost by each atom in the compound assuming the compound to be ionic (regardless of whether it is ionic or covalent). Elements have an oxidation number of 0 and thus the concept of oxidation numbers differs somewhat from that of valency. It is an artificial concept but a very useful one.

10.4 Extraction of the First Transition Series Elements

Scandium is of no significance and will not be considered. The majority of the other first transition series elements occur as oxides and with the exception of titanium the method of extraction employs chemical reduction of the appropriate oxide. Iron and nickel are extracted by reduction with carbon monoxide, while vanadium, chromium, manganese and cobalt are obtained by reduction of their oxides with aluminium. Titanium is obtained by reduction of titanium(IV) chloride, $TiCl_4$, with sodium, the chloride being first obtained from titanium(IV) oxide, TiO_2.

10.5 Some Reactions of Transition Metals and their Salts

(a) Action of acids on the metals

(i) *Action of hydrochloric acid*

Treat separate small samples (about 0·5 g) of as many of the first transition series metals as are available with approximately 2M hydrochloric acid and look for the evolution of gas. If there is no reaction in the cold try warming. If there is still no reaction use concentrated hydrochloric acid. Record your results and in particular make a note of the colours of the solutions so formed.

(ii) *Action of dilute nitric acid*

Repeat the experiment above using approximately 2M nitric acid in place of the hydrochloric acid, warming if necessary. Record all your observations, particularly the colours of any solutions formed.

(b) Reactions of some transition metal salts in aqueous solution

Use aqueous solutions of chromium(III) chloride, manganese(II) sulphate, ammonium iron(II) sulphate, iron(III) chloride, cobalt(II) chloride, nickel chloride and copper(II) sulphate, in which the concentration of the particular cation is approximately $0·1 \, mol \, dm^{-3}$, for the following experiments.

(i) *Action of sodium hydroxide solution*

Take about 2 cm^3 of each solution in separate test-tubes and then add, drop by drop, some approximately 2M sodium hydroxide solution. Note the colours of any precipitates and also observe which of the precipitates dissolve in an excess of the sodium hydroxide solution. Which, if any, of the hydroxides are amphoteric?

(ii) *Action of aqueous ammonia*

Repeat the above experiment, but replace the sodium hydroxide solution with approximately 2M aqueous ammonia. Again note the colours of the hydroxide precipitates, and also observe which of them dissolve in an excess of aqueous ammonia to give complex ions of the type $[M(NH_3)_x]^{n+}$.

(iii) *Action of hydrogen sulphide*

Take about 3 cm^3 of each solution in separate test-tubes and pass hydrogen sulphide through each **(in a fume cupboard)**. Note which solutions form precipitates and the colour of the precipitates. If no precipitate forms, make the solution alkaline with a little aqueous ammonia and note which solutions form a sulphide precipitate in alkaline solution. The precipitate which forms with the chromium(III) salt is chromium(III) hydroxide, i.e. chromium(III) sulphide does not exist in aqueous solution; can you explain why? The precipitate which forms with the iron(III) chloride is sulphur; can you write a balanced equation for the action of hydrogen sulphide on iron(III) chloride solution?

Now see which of the sulphide precipitates dissolve in hydrochloric acid. Try approximately 2M hydrochloric acid first, warming if necessary, and then concentrated hydrochloric acid if need be.

(iv) *Action of ammonium thiocyanate solution*

Take about 2 cm^3 of each solution in separate test-tubes and then add about the same volume of an approximately 0·1M solution of ammonium thiocyanate solution to each. Note any colour changes by comparison with a blank, i.e. a solution containing about 2 cm^3 of the transition metal salt solution and 2 cm^3 of distilled water in place of the ammonium thiocyanate solution. Which cations form complex ions with the thiocyanate ion, CNS^-?

Any coloration with the ammonium iron(II) sulphate solution is due to the presence of some iron(III) ions.

10.6 The Colours of Transition Metal Salts in Solution

You will have noticed from the experiments in Section 10.5(b) that transition metal salts are coloured in aqueous solution, and that the colour of a particular solution may be changed when water molecules are replaced by another ligand (atoms, ions or groups) in the complex, e.g. aqueous copper ions are light blue, but the colour deepens when water molecules are replaced by ammonia to give the $[Cu(NH_3)_4]^{2+}$ ion. Similarly, aqueous iron(III) ions are yellow, but a dark red colour is obtained on the addition of ammonium thiocyanate solution, due to the formation of the $[Fe(CNS)]^{2-}$ ion. Another colour change is given below.

Displacement of water molecules by chloride ions in the aqueous cobalt(II) ion

To about 3 cm^3 of a solution of cobalt(II) chloride add concentrated hydrochloric acid drop by drop and notice the final colour. Dilute with water and note the result, and again add some concentrated hydrochloric acid.

The colour changes in this example can be explained as being due to the replacement of water molecules in the pink $[Co(H_2O)_6]^{2+}$ ions by chloride ions to give the blue $[CoCl_4]^{2-}$ ions (potassium chloride would work just as well as hydrochloric acid):

$$[Co(H_2O)_6]^{2+} + 4Cl^- \rightleftharpoons [CoCl_4]^{2-} + 6H_2O$$

The colour of a transition metal ion is associated with:
(a) An incomplete d level (between 1 and 9 d electrons).
(b) The nature of the ligands surrounding the ion.

A complete theory of colour is very complex but, put simply, it is due to the movement of electrons from one d level to another. Since the five separate $3d$ orbitals are orientated differently in space, an electron (or electrons) which is close to a ligand will be repelled, and hence the energy of such orbitals will be raised relative to the others. The degeneracy of the $3d$ levels is therefore destroyed; this is represented pictorially for the copper(II) ion (Fig. 10.1).

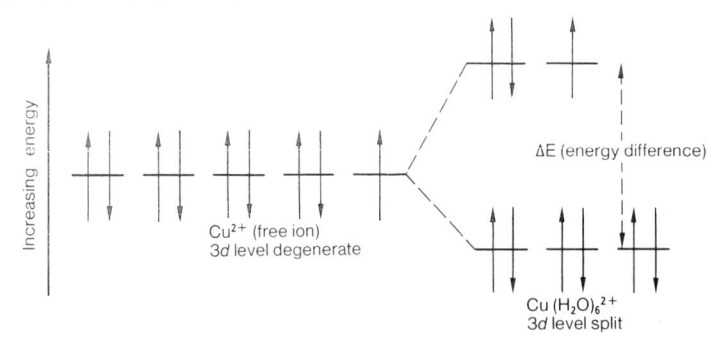

Fig. 10.1
The splitting of the $3d$ levels of the Cu^{2+} ion as a result of complex formation.

It will be seen from the above diagram that two of the $3d$ levels are raised in energy relative to the other three, the energy difference being $\triangle E$, where, by the Planck equation,

$$\triangle E = hv$$

in which h is Planck's constant and v is the frequency of light absorbed. Radiation of frequency given by the above equation will raise an electron from the lower $3d$ level to the upper one, and in the case of transition metal ions this radiation is part of the visible spectrum. Hydrated copper(II) ions are blue because yellow light of the appropriate frequency is absorbed (white light minus yellow light gives blue light). Since the degree of splitting of the $3d$ levels depends upon the particular ligands themselves, the variation in colour of ions of a particular transition metal is explained.

10.7 Complex Ions Formed by some Transition Metal Ions

A complex ion is one that contains a central ion linked to other atoms, ions or molecules which are called ligands. If the ligands are easily removed, the complex is said to be unstable, and if they are difficult to remove, the complex is a stable one. Typical complex ions include $[CoCl_4]^{2-}$, $[Cu(NH_3)_4]^{2+}$ and $[Fe(CNS)]^{2+}$. The stability of a complex ion can be discussed in a quantitative manner in terms of equilibrium reactions. Thus the addition of aqueous ammonia to an aqueous solution of a copper(II) salt results in the stepwise replacement of water molecules in the complex by ammonia molecules:

$$[Cu(H_2O)_4]^{2+} + NH_3 \rightleftharpoons [Cu(H_2O)_3NH_3]^{2+} + H_2O$$

$$[Cu(H_2O)_3NH_3]^{2+} + NH_3 \rightleftharpoons [Cu(H_2O)_2(NH_3)_2]^{2+} + H_2O$$

$$[Cu(H_2O)_2(NH_3)_2]^{2+} + NH_3 \rightleftharpoons [Cu(H_2O)(NH_3)_3]^{2+} + H_2O$$

$$[Cu(H_2O)(NH_3)_3]^{2+} + NH_3 \rightleftharpoons [Cu(NH_3)_4]^{2+} + H_2O$$

The equilibrium constant (stability constant) for the first reaction is given by the expression:

$$K_1 = \frac{[(Cu(H_2O)_3NH_3)^{2+}][H_2O]}{[(Cu(H_2O)_4)^{2+}][NH_3]} = 1 \cdot 1 \times 10^5 \text{ (at 293K)}$$

The stability constants for the other consecutive reactions, K_2, K_3 and K_4 are respectively $2 \cdot 3 \times 10^4$, $5 \cdot 3 \times 10^3$ and $9 \cdot 4 \times 10^2$. The overall stability constant K for the reaction:

$$[Cu(H_2O)_4]^{2+} + 4NH_3 \rightleftharpoons [Cu(NH_3)_4]^{2+} + 4H_2O$$

is the product $K_1 K_2 K_3 K_4 = 1 \cdot 3 \times 10^{16}$, which may easily be verified by simple mathematical manipulation of the equations above. The high value for this stability constant means that the $[Cu(NH_3)_4]^{2+}$ complex ion is far more stable than the simple hydrated complex ion $[Cu(H_2O)_4]^{2+}$.

In the examples so far mentioned the ligands have been attached to the central transition metal ion by one bond per ligand atom or molecule, such ligands being known as monodentate. However, there are many ligands which bond using two or more atoms per ligand molecule and are known as chelating agents; for example, the 2-hydroxybenzoate ion and the oxalate ion bond through the oxygen atoms marked with asterisks and are thus bidentate.

2-hydroxybenzoate ion   oxalate ion

Generally speaking, polydentate ligands form stabler complexes with a given transition metal ion than monodentate ones, although there is no hard and fast rule about this.

(a) Some experiments with complex ions using monodentate and bidentate ligands

(i) *The oxalate complex of chromium(III)*

Place about 3 cm^3 of a dilute solution of chromium(III) chloride in a test-tube and then add a small portion of solid sodium oxalate and shake the mixture. Note any change in colour of the solution. See if the resulting solution will form any chromium(III) hydroxide precipitate by adding approximately 2M sodium hydroxide solution dropwise. What do you conclude from your results?

$$Cr^{3+}(aq) + 3C_2O_4^{2-} \rightleftharpoons [Cr(C_2O_4)_3]^{3-} + water$$

(ii) *The oxalate complex of iron(III)*

Place about 3 cm^3 of a dilute solution of iron(III) chloride in a test-tube and then add two drops of ammonium thiocyanate solution to obtain a deep red-coloured solution containing the [Fe(CNS)]$^{2+}$ complex ion. Now add, in small portions with shaking, some solid sodium oxalate and note the final colour of the solution. Will the final solution produce a red colour on the addition of more ammonium thiocyanate solution? Explain all your observations in terms of the equilibria:

$$Fe^{3+}(aq) + CNS^- \rightleftharpoons [Fe(CNS)]^{2+} + water$$
$$Fe^{3+}(aq) + 3C_2O_4^{2-} \rightleftharpoons [Fe(C_2O_4)_3]^{3-} + water$$

(iii) *The fluoride complex of iron(III)*

Place about 3 cm^3 of a dilute solution of iron(III) chloride in a test-tube and then add a small portion of solid sodium fluoride. What happens to the colour of the solution as the [FeF$_6$]$^{3-}$ complex ion is formed?

$$Fe^{3+}(aq) + 6F^- \rightleftharpoons [FeF_6]^{3-} + \text{water}$$

Divide the resulting solution into two equal portions. To one portion add a little ammonium thiocyanate solution and note whether the red colour, which confirms the presence of $Fe^{3+}(aq)$ ions, appears. To the other portion add some approximately 2M sodium hydroxide solution and observe whether there are sufficient $Fe^{3+}(aq)$ ions present to exceed the solubility product of iron(III) hydroxide. Note and explain both sets of results.

(iv) *Examination of the equilibrium:* $2Fe^{3+} + 2I^- \rightleftharpoons 2Fe^{2+} + I_2$

Place about $3\,cm^3$ of a dilute solution of iron(III) chloride in a test-tube and add a little potassium iodide solution. Is any iodine liberated? Add a little tetrachloromethane if you are in any doubt, and see whether the tetrachloromethane layer becomes purple on shaking. Repeat the experiment, but, before adding the potassium iodide solution, add a small portion of solid sodium fluoride to the iron(III) chloride solution. Are the two results similar or different?

Place about $3\,cm^3$ of a dilute solution of iron(II) sulphate in a test-tube and add a little dilute iodine solution (iodine dissolved in potassium iodide solution). Warm, but do not heat so strongly that the iodine vaporises. Is there any noticeable reaction? Repeat the experiment, but add a small portion of solid sodium fluoride to the iron(II) sulphate solution before adding the iodine solution. Again, warm gently and see if there is any noticeable reaction.

Explain all your results obtained in this experiment in terms of the two equilibrium reactions:

$$2Fe^{3+}(aq) + 2I^- \rightleftharpoons 2Fe^{2+}(aq) + I_2$$
$$2[FeF_6]^{3-} + 2I^- \rightleftharpoons 2[FeF_6]^{4-} + I_2$$

Are your results consistent with the following standard redox potentials? Explain fully.

$$2Fe^{3+}(aq) + 2e^- \rightarrow 2Fe^{2+}(aq) \qquad E^\ominus = +0.76\,V$$
$$2[FeF_6]^{3-} + 2e^- \rightarrow 2[FeF_6]^{4-} \qquad E^\ominus = +0.4\,V$$
$$I_2 + 2e^- \rightarrow 2I^- \qquad E^\ominus = +0.54\,V$$

(b) Some experiments with complex ions using ethylenediaminetetra-acetic acid (EDTA)—a hexadentate ligand

Ethylenediaminetetra-acetic acid (EDTA) has the formula

$$\begin{array}{c} HOOC-H_2C \\ \diagdown \\ N-CH_2-CH_2-N \\ \diagup \\ HOOC-H_2C \end{array} \begin{array}{c} CH_2-COOH \\ \diagup \\ \\ \diagdown \\ CH_2-COOH \end{array}$$

and is a tetrabasic acid but, since it is not very soluble in water, it is usual to use the disodium salt. Under appropriate conditions the quadruply charged anion shown below can be obtained; this anion forms 1:1 complexes of great stability with many metallic cations (1 mole of the anion reacts with 1 mole of the cation), co-ordinating on to the cations by means of lone pairs of electrons (marked with an asterisk below).

$$\begin{array}{c} -\overset{*}{O}OC-H_2C \\ \diagdown \\ -\overset{*}{O}OC-H_2C \end{array} \overset{*}{N}-CH_2-CH_2-\overset{*}{N} \begin{array}{c} CH_2-CO\overset{*}{O}- \\ \diagup \\ CH_2-CO\overset{*}{O}- \end{array}$$

In order to obtain the EDTA completely in the form of the above anion, it is necessary to add OH^- ions, which remove the H^+ ions as unionised water molecules, and thus allow the following four equilibria to swing over completely to the right (it is customary to denote EDTA as H_4Y when discussing these reactions).

$$H_4Y \rightleftharpoons H_3Y^- + H^+$$
$$H_3Y^- \rightleftharpoons H_2Y^{2-} + H^+$$
$$H_2Y^{2-} \rightleftharpoons HY^{3-} + H^+$$
$$HY^{3-} \rightleftharpoons Y^{4-} + H^+$$

(i) *Hydrogen ions are released when EDTA complexes with an aqueous solution of a cobalt(II) salt*

Place about 3 cm^3 of a dilute solution of cobalt(II) nitrate in a test-tube and add two drops of B.D.H. '4.5' indicator solution. Note the colour of the solution and then add about 3 cm^3 of a dilute aqueous solution of EDTA. Note the colour change of the indicator. Are your results consistent with a reaction such as:

$$Co^{2+} + EDTA \rightarrow Co(EDTA) + 2H^+ ?$$

(ii) *The colours of the EDTA complexes of* Co^{2+}, Cu^{2+}, Fe^{2+} *and* Ni^{2+}

Place about 3 cm^3 of a dilute solution of cobalt(II) nitrate in each of two separate test-tubes and add a small portion of solid EDTA to one of the test-tubes. Compare the colours of the two solutions, adding a little more EDTA if there is not much difference. Repeat the experiment using aqueous solutions of copper(II) sulphate, iron(II) sulphate and nickel sulphate in place of the cobalt(II) nitrate. Record the colours of these EDTA complexes.

(iii) *Displacement of the 2-hydroxybenzoate ion from an iron(III) complex with EDTA*

To about 3 cm^3 of a dilute solution of iron(III) chloride in propanone (acetone) add about the same volume of a dilute solution of 2-hydroxybenzoic acid (salicylic acid) also in propanone. Note the colour of the iron(III)-

2-hydroxybenzoate complex (the 2-hydroxybenzoate ion is bidentate and its structure is given on p. 154). Now add a small portion of EDTA and shake, adding a little more EDTA if there is no significant change in colour. Record and explain your observations.

(iv) *The reaction between aqueous solutions containing* Fe^{2+} *and* Ag^+ *in the presence of EDTA*

Add about six drops of silver nitrate solution to about $3\,cm^3$ of a dilute solution of iron(II) sulphate (or better, ammonium iron(II) sulphate). Can you detect any evidence of reaction? Now add a small portion of solid EDTA and shake. Record what you observe and use the following redox potentials to explain your observations:

$$Fe^{3+}(aq) + e^- \rightarrow Fe^{2+}(aq) \qquad E^\ominus = +0\cdot76\,V$$
$$Fe^{3+}(EDTA) + e^- \rightarrow Fe^{2+}(EDTA) \qquad E^\ominus = +0\cdot12\,V$$
$$Ag^+(aq) + e^- \rightarrow Ag \qquad E^\ominus = +0\cdot80\,V$$

10.8 Variable Oxidations States

Except for scandium which has exclusively an oxidation state of $+3$, the first transition series elements show an oxidation state of $+2$ when both $4s$ electrons are involved in bonding; for oxidation states greater than $+2$, $3d$ electrons are used in addition to both $4s$ electrons.

(a) Conversion of Fe(II) to Fe(III) and vice versa

Carry out Experiment 6.7(f)(i) p. 79 and Experiment 8.12(a)(iii) p. 121, if you have not previously done them. In the first case Fe^{2+} is oxidised to Fe^{3+}, and in the second Fe^{3+} is reduced to Fe^{2+}.

(b) Stepwise conversion of vanadium(V) to vanadium(II) and oxidation of vanadium(II) with concentrated nitric acid

Place about $10\,cm^3$ of ammonium metavanadate solution, $NH_4^+VO_3^-$ in a boiling-tube and then add a few pieces of granulated zinc and $5\,cm^3$ of concentrated hydrochloric acid. Leave the mixture in a rack and observe all the colour changes (the first green colour is due to a mixture of vanadium(V) and vanadium(IV) and should be ignored). Record the colours of the various oxidation states below.

Ion	VO_3^- Metavanadate	VO^{2+} Vanadium(IV)	V^{3+} Vanadium(III)	V^{2+} Vanadium(II)
Oxidation state	$+5$	$+4$	$+3$	$+2$
Colour				

After the final colour change has occurred, pour about 5 cm³ of the solution into a test-tube and then add, with shaking, some concentrated nitric acid, drop by drop. To which oxidation state does nitric acid convert vanadium(II)?

Are your results consistent with the following standard redox potentials?

$$Zn^{2+} + 2e^- \rightarrow Zn \qquad E^\ominus = -0.76 \text{ V}$$
$$VO_3^- + 4H^+ + e^- \rightarrow VO^{2+} + 2H_2O \qquad E^\ominus = +1.00 \text{ V}$$
$$VO^{2+} + 2H^+ + e^- \rightarrow V^{3+} + H_2O \qquad E^\ominus = +0.34 \text{ V}$$
$$V^{3+} + e^- \rightarrow V^{2+} \qquad E^\ominus = -0.26 \text{ V}$$
$$2NO_3^- + 4H^+ + 2e^- \rightarrow 2NO_2 + 2H_2O \qquad E^\ominus = +0.81 \text{ V}$$

(c) Conversion of chromium(VI) to chromium(II) and oxidation of chromium(II) with concentrated nitric acid

Place about 10 cm³ of a dilute solution of potassium dichromate, $(K^+)_2Cr_2O_7^{2-}$, in a boiling-tube and then add a few pieces of granulated zinc and 5 cm³ of concentrated hydrochloric acid. Warm slightly until reduction is underway and then leave the tube in a rack and observe the colour changes as below:

Ion	$Cr_2O_7^{2-}$ Dichromate	Cr^{3+} Chromium(III)	Cr^{2+} Chromium(II)
Oxidation state	+6	+3	+2
Colour			

After the final colour change has occurred, pour about 5 cm³ of the solution into a test-tube and then add, with shaking, some concentrated nitric acid, drop by drop. To which oxidation state do you think the nitric acid converts chromium(II)?

Are your results consistent with the following standard redox potentials?

$$Zn^{2+} + 2e^- \rightarrow Zn \qquad E^\ominus = -0.76 \text{ V}$$
$$Cr_2O_7^{2-} + 14H^+ + 6e^- \rightarrow 2Cr^{3+} + 7H_2O \qquad E^\ominus = +1.33 \text{ V}$$
$$Cr^{3+} + e^- \rightarrow Cr^{2+} \qquad E^\ominus = -0.41 \text{ V}$$
$$2NO_3^- + 4H^+ + 2e^- \rightarrow 2NO_2 + 2H_2O \qquad E^\ominus = +0.81 \text{ V}$$

10.9 Catalytic Activity

The catalytic activity of transition metals and their compounds is associated with their variable oxidation states. Typical catalysts are vanadium(V) oxide used in the Contact Process for making sulphuric acid, finely divided iron in the Haber process for making ammonia, and nickel in the hydrogenation of oils to fats, e.g. margarine.

Catalysis at a solid surface involves the formation of bonds between reactant molecules and the surface catalyst atoms (the first transition series elements have $3d$ electrons in addition to the $4s$ electrons, which can be utilised in bonding); this has the effect of increasing the concentration of the reactant molecules at the catalyst surface and also of weakening the bonds in the reactant molecules (the activation energy is lowered).

Transition metal ions function as catalysts by changing their oxidation states, substituting two or more easier reaction paths for one more difficult one.

(a) The catalysed and uncatalysed reaction between iodide ions and persulphate ions in aqueous solution

The reaction studied is that between iodide ions and persulphate ions in the presence of a little sodium thiosulphate solution. Iodide ions reduce persulphate ions to sulphate ions, and are themselves oxidised to iodine by a slow reaction. However, in the presence of thiosulphate ions, the iodine formed is used up in a fast reaction, and does not appear until all the thiosulphate has been consumed. Its colour is intensified by the addition of starch solution, and the time taken for a dark colour to appear indicates when a definite fraction of the iodide/persulphate reaction is complete.

$$2I^- + S_2O_8^{2-} \rightarrow I_2 + 2SO_4^{2-}$$

Place 10 cm^3 of 0.2M potassium iodide solution, 5 cm^3 of 0.01M sodium thiosulphate solution and 5 cm^3 of a 5% starch solution in a conical flask. Now add 20 cm^3 of a saturated solution of potassium persulphate and immediately start a stop watch. Note the time when a dark colour, due to the formation of iodine, appears.

Repeat the experiment as above, but this time add five drops of a 0.1M solution of Fe^{3+} to the 20 cm^3 solution of persulphate solution, before it is added to the iodide solution. Again, note the time taken for a dark colour to appear. Do Fe^{3+} ions catalyse the reaction? Given the redox potentials below, can you write two reactions which might explain this catalysis?

$$S_2O_8^{2-} + 2e^- \rightarrow 2SO_4^{2-} \qquad E^\ominus = +2 \cdot 01 \text{ V}$$
$$I_2 + 2e^- \rightarrow 2I^- \qquad E^\ominus = +0 \cdot 54 \text{ V}$$
$$Fe^{3+} + e^- \rightarrow Fe^{2+} \qquad E^\ominus = +0 \cdot 76 \text{ V}$$

Would you expect Fe^{2+} ions to catalyse the reaction? Try the experiment using a 0.1M solution of Fe^{2+} (made from ammonium iron(II) sulphate) in place of the Fe^{3+}.

Given that the standard redox potential for the system $Co^{3+} + e^- \rightarrow Co^{2+}$ is $+1 \cdot 82$ V, decide whether it is theoretically possible for Co^{2+} ions to catalyse the reaction. Try the experiment using a 0.1M solution of Co^{2+} in place of the Fe^{3+}. What explanation can you give of your observations?

(b) The reaction between tartrate ions and hydrogen peroxide catalysed by cobalt(II) ions—the formation of an intermediate compound

Place 5 cm^3 of an approximately 0.4M solution of sodium potassium tartrate, $Na^+K^+C_4H_4O_6^{2-} \cdot 4H_2O$, in a boiling-tube and then add 5 cm^3 of

'20 vol' hydrogen peroxide solution. Warm the solution until it is gently boiling and then allow it to cool somewhat. Is there much evidence of gas evolution? Now add a little cobalt(II) chloride solution until the mixture is pink and then stand the tube in a rack. Note your observations, in particular the colour changes of the solution. The gas evolved is carbon dioxide with some oxygen.

Repeat the experiment, but immediately place the boiling-tube in a beaker containing ice-cold water when gas evolution commences. You should be able to slow down the reaction to such an extent that it is possible to preserve the intermediate compound for some considerable time.

The overall reaction may be described as:

$$\begin{array}{c} COO^- \\ | \\ H-C-OH \\ | \\ H-C-OH \\ | \\ COO^- \end{array} + 2H_2O \rightarrow 4CO_2 + 8H^+ + 10e^-$$

$$2H^+ + H_2O_2 + 2e^- \rightarrow 2H_2O$$

or $\quad C_4H_4O_6^{2-} + 5H_2O_2 + 2H^+ \rightarrow 4CO_2 + 8H_2O$

Can you suggest how cobalt(II) ions might catalyse this reaction?

10.10 Paramagnetism

Many transition metal compounds are weakly attracted by a strong non-homogeneous magnetic field, the phenomenon being referred to as paramagnetism. Paramagnetism arises because electrons can be regarded as spinning on their axes and, just as an electric current flowing through a wire generates a magnetic moment, so too does a spinning electron; electrons of course that occupy the same orbital have zero magnetic moment, since the two electrons involved will be spinning in opposite directions. The manganese(II) ion, for example, has five unpaired $3d$ electrons and should thus give the largest paramagnetic effect observed in the first transition series:

Mn $3d^5 4s^2$ $\quad$ Mn^{2+} $3d^5$ $\quad$ |↑|↑|↑|↑|↑|

To demonstrate the paramagnetism of manganese(II) sulphate

A small plastic container (the barrel of a plastic syringe about 8 cm long and 0·5 cm in diameter) is sealed at one end with a small plastic cap and filled with finely divided manganese(II) sulphate. The container is now suspended between the poles of a powerful electromagnet, capable of producing a very intense non-homogeneous magnetic field (see Plate 2). When the current is switched on, the container should be attracted to one of the poles of the magnet. Show that the empty container is not itself attracted by the magnet, and also that a substance that contains no unpaired electrons, for example sodium chloride, shows no effect.

161

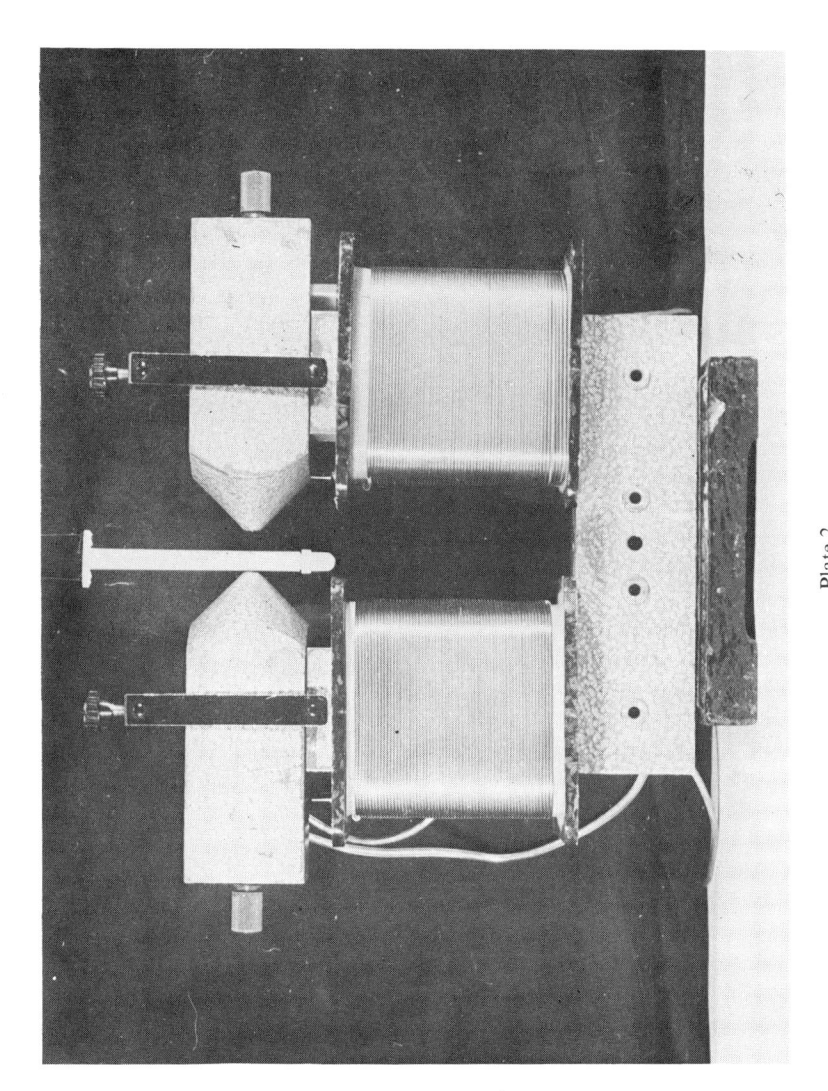

Plate 2
The apparatus used to demonstrate the paramagnetism of manganese(II) sulphate.

11

THE FIRST TRANSITION SERIES IN MORE DETAIL (SCANDIUM, TITANIUM, VANADIUM, CHROMIUM, MANGANESE, IRON, COBALT, NICKEL)

11.1 Vanadium

(a) Survey of the oxidation states of vanadium

For this experiment you will require a solution of vanadium(V) containing about 0·05 mole of vanadium(V) per 1000 cm^3 of solution. This may be obtained by warming about 5 g of vanadium(V) oxide, V_2O_5, and 8 g of sodium hydroxide with about 40 cm^3 of distilled water, until a solution is obtained. The solution should be just acidified with approximately M sulphuric acid and made up to 1000 cm^3 with distilled water. The exact concentration of this solution is not required. A solution of approximately 0·02M potassium permanganate (exact concentration is not required) is also needed.

Pipette 25 cm^3 of the vanadium(V) solution into a conical flask and then add 25 cm^3 of distilled water, 50 cm^3 of approximately 2M sulphuric acid and 3 g of granulated zinc. Heat the solution until the reduction is complete (about 30 minutes) and then pour the solution, while still boiling, through a little glass wool into another flask containing exactly 25 cm^3 of the 0·02M potassium permanganate solution. Rinse the conical flask with hot boiled-out distilled water, pouring the washings through the glass wool, and complete the titration at 70–80°C, the end-point being when a faint trace of pink colour is obtained.

Repeat the experiment exactly as above, but this time filter the solution into an empty Buchner flask and again rinse the conical flask with hot distilled water, pouring the washings through the glass wool. Fit the Buchner flask with a rubber bung carrying a glass tube and then suck air through the solution, using a water pump, for about five minutes. Allow the solution to stand for about 30 minutes and then titrate with 0·02M potassium permanganate solution at 70–80°C as before.

Pipette 25 cm^3 of the vanadium(V) solution into a conical flask and then add 25 cm^3 of approximately 2M sulphuric acid and 50 cm^3 of distilled water. Pass sulphur dioxide through the solution for about five minutes. Boil to expel excess sulphur dioxide, cool to 70–80°C and titrate with 0·02M potassium permanganate solution.

Potassium permanganate oxidises the various oxidation states of vanadium back to vanadium(V), hence it is possible to determine the other oxidation states of the metal from the titration readings. What are they?

11.2 Chromium

Chromium has oxidation states of $+6$ (oxidising), $+3$ (the most stable) and $+2$ (reducing). These have been demonstrated previously by the reduction of a solution of potassium dichromate with zinc in the presence of hydrochloric acid (see Experiment 10.8(c), p. 158).

(a) Chromium(VI) compounds

(i) *Reduction of potassium dichromate solution with an iron(II) solution — a quantitative experiment*

Potassium dichromate, $(K^+)_2Cr_2O_7^{2-}$, unlike potassium permanganate, can be used as a primary standard since it can be obtained in a pure state and its solution is stable and keeps indefinitely. It is used mainly for the titration of iron(II) salts.

During a dichromate titration (which is done in acid solution) the orange $Cr_2O_7^{2-}$ ions are converted into green Cr^{n+} ions; thus, unlike potassium permanganate, it cannot act as its own indicator. The indicator used is barium diphenylamine sulphonate, which changes colour at the end-point to blue/purple; since the iron(III) ions produced during the titration also oxidise the indicator, they must be removed as fast as they are formed, and this is done by adding phosphoric acid with which they form a stable complex.

The object of the experiment is to determine the number of moles of iron(II) ions that react with one mole of dichromate ions and hence to construct the stoichiometric equation for this reaction. Having done this, the value of n in Cr^{n+} can be determined. The only additional information you need is that iron(II) ions are oxidised to iron(III) ions and that the added acid (H^+ ions) reacts with the oxygen atoms in the dichromate ions to form water.

Using a dropping tube, place 25 drops of a 0·1M solution of ammonium iron(II) sulphate in a test-tube, add about the same volume of approximately 2M sulphuric acid, 2 drops of a 0·2% solution of barium diphenylamine sulphonate indicator and then 5 drops of phosphoric acid (50% syrupy phosphoric acid and 50% water by volume). Using the same clean dropping tube add, drop by drop with shaking, a 0·018M solution of potassium dichromate until the indicator changes sharply to blue/violet (you should be able to determine the volume to the nearest drop). Now calculate, to the nearest whole number, the number of moles of iron(II) ions that react with one mole of dichromate ions. Having done this, construct the stoichiometric equation for the reaction between dichromate and iron(II) ions in the presence of H^+ ions and hence determine the value of n in Cr^{n+}.

(ii) *The action of acid and alkali on aqueous solutions of potassium dichromate and potassium chromate*

The dichromate ion, $Cr_2O_7^{2-}$, is orange and the chromate ion, CrO_4^{2-}, is yellow. Both ions contain chromium in the oxidation state $+6$.

Place about 3 cm³ of an aqueous solution of potassium dichromate in a test-tube and then add some approximately 2M sodium hydroxide solution until a colour change occurs. Now acidify with approximately M sulphuric acid and observe what happens. Repeat the experiment with about 3 cm³ of an aqueous solution of potassium chromate, but acidify first, and then, after a colour change has occurred, make the solution alkaline with some sodium hydroxide solution. Attempt to write a balanced ionic equation to show the effect of H^+ and OH^- ions on the dichromate/chromate equilibrium.

(iii) *Formation of chromyl chloride (chromium(VI) dichloride dioxide), CrO_2Cl_2, and its reaction with alkali*

Place about 0·3 g of sodium chloride in a test-tube and mix it with about the same volume of potassium dichromate. Add several drops of concentrated sulphuric acid and warm gently **(care)**. Note the colour of the chromyl chloride vapour.

$$Cr_2O_7^{2-} + 4Cl^- + 6H^+ \rightarrow 2CrO_2Cl_2 + 3H_2O$$

Fit a delivery tube to the test-tube and pass the chromyl chloride vapour into a test-tube containing about 2 cm³ of approximately 2M sodium hydroxide solution and note the colour change. **Caution: do not allow the alkali to suck back into the hot acid solution.** Attempt to write an equation to describe the action of alkali on chromyl chloride.

(b) Chromium(III) compounds

(i) *Preparation of potassium chromium(III) sulphate, $K^+Cr^{3+}(SO_4^{2-})_2 \cdot 12H_2O$*

To a solution of 5 g of potassium dichromate in 40 cm³ of water add, cautiously, 1 cm³ of concentrated sulphuric acid. Cool the solution and then pass sulphur dioxide into it until a greenish-blue solution is obtained (**this operation should be done in a fume cupboard**). Pour the solution into an evaporating dish and allow it to crystallise at room temperature.

$$2K^+ + Cr_2O_7^{2-} + 2H^+ + SO_4^{2-} + 3SO_2 + 23H_2O \rightarrow$$
$$2K^+Cr^{3+}(SO_4^{2-})_2 \cdot 12H_2O$$

(ii) *Precipitation of the sulphate ions in potassium chromium(III) sulphate by lead(II) nitrate solution under different conditions*

Prepare a 0·1M solution of potassium chromium(III) sulphate by dissolving 5 g of the solid in cold water (**do not heat**) making the total volume up to 100 cm³ in a standard flask. Note the colour of this solution. Pour about half of the solution into a small beaker and warm to a temperature above 70°C until the solution changes colour. Note the colour of this solution. Allow it to cool down to room temperature. Does the former colour return?

Place 5 cm³ of each of the two solutions into two separate 10 cm³ graduated cylinders and add 5 cm³ of 0·2M lead(II) nitrate solution to each. Stir each mixture with a glass rod and then allow the white precipitates of lead(II) sulphate to settle. Does each mixture produce the same amount of lead(II) sulphate precipitate? Try to make an estimate of the relative amounts of sulphate precipitate in each case.

Given that the colour of the first solution of potassium chromium(III) sulphate, i.e. the one not heated, is due to the $[Cr(H_2O)_6]^{3+}$ ion and using your results above, suggest what complex ion might be the cause of the colour in the second solution of potassium chromium(III) sulphate.

(c) Chromium(II) compounds

Reduction of potassium dichromate solution with zinc and hydrochloric acid

An acidified solution of potassium dichromate can be reduced to a blue solution containing Cr^{2+}(aq) with zinc and hydrochloric acid. Carry out Experiment 10.8(c), p. 158, if you have not already done it.

Is this result consistent with the following standard redox potentials?

$$Cr_2O_7^{2-} + 14H^+ + 6e^- \rightarrow Cr^{3+} + 7H_2O \qquad E^\ominus = +1\cdot33\text{ V}$$
$$Cr^{3+} + e^- \rightarrow Cr^{2+} \qquad E^\ominus = -0\cdot41\text{ V}$$
$$Zn^{2+} + 2e^- \rightarrow Zn \qquad E^\ominus = -0\cdot76\text{ V}$$

What would you expect to happen when an acidified solution of potassium dichromate is treated with an aqueous solution of an iron(II) salt, given the standard redox potential?

$$Fe^{3+} + e^- \rightarrow Fe^{2+} \qquad E^\ominus = +0\cdot76\text{ V}$$

Is your answer consistent with the result of Experiment 11.2(a)(i), p. 163?

11.3 Manganese

Manganese has oxidation states of $+7$ (powerfully oxidising), $+6$, $+4$, $+3$ and $+2$ (the most stable).

(a) Manganese(VII) compounds

(i) *Preparation of potassium permanganate (potassium manganate(VII))*

Place 10 g of potassium hydroxide in an iron dish and warm gently until it has melted (**caution**). Now add 1 g of potassium chlorate(V), mixing it into the melt with a glass rod, and then 7·5 g of manganese(IV) oxide, in portions, continuing to stir the mixture. When all the manganese(IV) oxide has been added, heat to bright red heat for about 15 minutes. Allow the mixture to cool and then grind it to a fine powder in a mortar. Transfer the powder to a 500 cm³ flask, add 200 cm³ of distilled water and heat, passing a stream of carbon dioxide into the solution. At first the solution is green (due to the presence of potassium manganate(VI)):

$$3MnO_2 + 6OH^- + ClO_3^- \rightarrow 3MnO_4^{2-} + 3H_2O + Cl^-$$
manganate(VI) green

Later, it becomes purple as potassium permanganate is formed by disproportionation of the potassium manganate(VI):

$$3MnO_4^{2-} + 2H_2O \rightleftharpoons 2MnO_4^- + MnO_2 + 4OH^-$$
permanganate

Can you explain how the carbon dioxide assists in moving the above equilibrium to the right?

After boiling for 10 minutes allow to cool and decant through glass wool. Wash the flask and return the filtrate to it. Boil for a further 10 minutes while carbon dioxide passes through, and then withdraw a drop of the solution and place it on a filter paper. If the drop has a green centre continue the process until complete. Finally cool and filter. Transfer the solution to a large evaporating dish and evaporate until crystals appear, then set aside to crystallise.

(ii) *Conversion of potassium permanganate to potassium manganate(VI)*

To about 1 cm³ of a saturated solution of potassium permanganate add about the same volume of a concentrated solution of potassium hydroxide. Warm **cautiously** and note the colour of the final solution which contains potassium manganate(VI). Identify the gas evolved. Attempt to write a balanced equation for the reaction between permanganate ions and hydroxyl ions.

(iii) *Reduction of potassium permanganate solution with an iron(II) solution—a quantitative experiment*

The object of this experiment is to determine the number of moles of iron(II) ions that react with an acidified solution of permanganate ions and hence to construct the stoichiometric equation for this reaction. During the reaction, MnO_4^- ions are converted into Mn^{n+} ions. The only additional information you need is that iron(II) ions are oxidised to iron(III) ions and that the added acid(H^+) reacts with the oxygen atoms in the permanganate ions to form water.

Using a dropping tube, place 25 drops of a 0·1M solution of ammonium iron(II) sulphate in a test-tube and add about the same volume of approximately 2M sulphuric acid. Using the same clean dropping tube add, drop by drop with shaking, a 0·018M solution of potassium permanganate until a permanent pink colour remains after shaking (you should be able to determine the volume to the nearest drop). Now calculate, to the nearest whole number, the number of moles of iron(II) ions that react with one mole of permanganate ions. Having done this, construct the stoichiometric equation for the reaction between permanganate and iron(II) ions in the presence of H^+ ions, and hence determine the value of n in Mn^{n+}.

(b) Manganese(VI) compounds

The only pure compounds so far obtained containing manganese in the +6 oxidation state are sodium and potassium manganates(VI); both compounds are dark green solids, the colour being due to the MnO_4^{2-} ion.

(i) *Formation of potassium manganate(VI)*

This is formed during the preparation of potassium permanganate (see Experiment 11.3(a)(i), p. 165). It can also be formed by the reaction between potassium permanganate and manganese(IV) oxide in strongly alkaline solution.

Place about 0·3 g of potassium permanganate and 0·3 g of manganese(IV) oxide in a test-tube and add 3 or 4 pellets of potassium hydroxide. **Warm very cautiously** and note any colour change, diluting with water if the final colour is not too apparent. Attempt to write a balanced ionic equation for the reaction between permanganate and manganese(IV) oxide in alkaline solution.

(ii) *Disproportionation of potassium manganate(VI)*

Dissolve about 0·3 g of potassium manganate(VI) in 3 cm³ of distilled water and add approximately M sulphuric acid, drop by drop. Note any colour changes. Potassium permanganate and manganese(IV) oxide are formed by reversal of the reaction (i) above. What is the purpose of the sulphuric acid?

Are the results of this experiment consistent with the following redox potentials?

$$2MnO_4^- + 2e^- \rightarrow 2MnO_4^{2-} \qquad E^\ominus = +0.56 \text{ V}$$
$$MnO_4^{2-} + 4H^+ + 2e^- \rightarrow MnO_2 + 2H_2O \qquad E^\ominus = +2.26 \text{ V}$$

Since potassium manganate(VI) can be formed from potassium permanganate and manganese(IV) oxide in strongly alkaline solution, what can you say about the redox potential for the system

$$MnO_4^{2-} + 2H_2O + 2e^- \rightarrow MnO_2 + 4OH^-$$

under these conditions?

(c) Manganese(IV) compounds

The only compound of manganese(IV) of any importance is manganese(IV) oxide, MnO_2, which occurs naturally as the ore pyrolusite. It is an oxidising agent.

(i) *Action of heat on manganese(IV) oxide*

Heat about 0·5 g of manganese(IV) oxide in an ignition tube and test for the evolution of oxygen with a glowing splint:

$$3MnO_2 \rightarrow Mn_3O_4 + O_2$$

(ii) *Action of concentrated hydrochloric acid on manganese(IV) oxide*

Place about 0·3 g of manganese(IV) oxide in a test-tube and add about 1 cm³ of concentrated hydrochloric acid. Test for the evolution of chlorine with moist blue litmus paper:

$$MnO_2 + 4HCl \rightarrow MnCl_2 + Cl_2 + 2H_2O$$

(d) Manganese(III) compounds

A solution containing manganese(III) can be obtained by the oxidation of manganese(II) with manganese(VII) (potassium permanganate) in strongly acid solution. This is demonstrated below, together with the preparation of the covalent compound tris (acetylacetonato) manganese(III).

(i) *Formation of manganese (III) in solution*

Place about 0·5 g of manganese(II) sulphate in a test-tube and dissolve it in about 2 cm³ of approximately M sulphuric acid. Add 10 drops of concentrated sulphuric acid and, after cooling to room temperature, 5 drops of 0·1M potassium permanganate solution. Compare the colour of this solution which contains manganese(III) with that obtained by adding 5 drops of 0·1M potassium permanganate solution to about 3 cm³ of distilled water (control test).

$$MnO_4^- + 4Mn^{2+} + 8H^+ \rightarrow 5Mn^{3+} + 4H_2O$$

The +3 oxidation state of manganese is here stabilised in the form of a complex sulphate.

Now pour the solution containing manganese(III) into about 50 cm³ of distilled water and stir. Record your observations.

(iii) *Preparation of tris(acetylacetonato) manganese(III),*
$Mn(CH_3COCHCOCH_3)_3$

Place 75 cm³ of 0·16M manganese(II) chloride solution in a 250 cm³ beaker and add 25 cm³ of approximately 2M sodium acetate solution. Now add 10 cm³ of acetylacetone, $CH_3COCH_2COCH_3$, stir thoroughly and then slowly run into the mixture 25 cm³ of 0·12M potassium permanganate solution. Finally, add 25 cm³ of approximately 2M sodium acetate solution and heat on a water bath for 10 minutes. Place the beaker containing the mixture in a larger beaker containing ice and water and allow the product to crystallise. Filter off the crystals of tris(acetylacetonato) manganese(III) and allow them to dry.

The overall reaction is:

$$4Mn^{2+} + MnO_4^- + 7CH_3COO^- + 15CH_3COCH_2COCH_3 \rightarrow$$
$$5Mn(CH_3COCHCOCH_3)_3 + 4H_2O + 7CH_3COOH$$

The enol-form of acetylacetone is $CH_3-\underset{\underset{OH}{|}}{C}=CH-\underset{\underset{O}{\|}}{C}-CH_3$ and three

enolate ions, $CH_3-\underset{\underset{O^-}{|}}{C}=CH-\underset{\underset{O}{\|}}{C}-CH_3$, are attached to the Mn^{3+} ion by

means of covalent bonds (see (opposite); the complex is thus neutral.

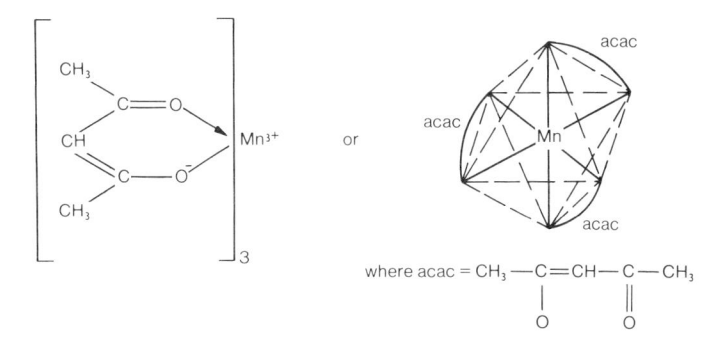

where acac = $CH_3-C=CH-C-CH_3$ with O below the first C and ‖O below the second C

(e) Manganese(II) compounds

This is the most stable oxidation state of manganese; the presence of five singly occupied $3d$ orbitals in the Mn^{2+} ion is often cited as an explanation of this stability.

The action of sodium hydroxide solution, aqueous ammonia and hydrogen sulphide on aqueous solutions of a manganese(II) salt has been dealt with in Section 10.5, p. 150. The oxidation of manganese(II) to manganese(VII) has previously been carried out in Section 6.16(c), p. 90.

11.4 Iron

Iron shows oxidation states of $+3$ (the most stable) and $+2$ (reducing). There is also an unstable oxidation state of $+6$ which is powerfully oxidising. The presence of five singly occupied $3d$ orbitals in the Fe^{3+} ion and only four in the Fe^{2+} ion is often cited as an explanation of the easy oxidation of Fe^{2+} to Fe^{3+}.

(a) Iron(VI) compounds

The ferrate(VI) ion, FeO_4^{2-}, contains iron in the $+6$ oxidation state. It is a powerful oxidising agent.

(i) *Preparation of barium ferrate(VI)*, $Ba^{2+}FeO_4^{2-}$

Place $50\,cm^3$ of approximately 2M sodium chlorate(I) solution in a $250\,cm^3$ beaker and dissolve in this solution 2 g of sodium hydroxide. Heat the mixture until it begins to boil and then add, drop by drop with stirring, $5\,cm^3$ of 0·5M iron(III) nitrate solution. When the addition is complete, boil for about 2 minutes and then filter the reddish-violet solution through glass wool into a conical flask. Now add a solution of barium chloride to precipitate barium ferrate(VI). Filter off the red solid, wash it with a little distilled water and then allow it to dry.

The reactions involved in the above preparation are:

$$2Fe^{3+} + 3OCl^- + 10OH^- \rightarrow 2FeO_4^{2-} + 3Cl^- + 5H_2O$$
$$Ba^{2+} + FeO_4^{2-} \rightarrow Ba^{2+}FeO_4^{2-}$$

The action of acid on a ferrate(VI) produces oxygen and the iron(VI) passes into iron(III):

$$4FeO_4^{2-} + 20H^+ \rightarrow 4Fe^{3+} + 3O_2 + 10H_2O$$

Try the effect of hydrochloric acid on a sample of your barium ferrate(VI).

(a) Iron(III) compounds and Iron(II) compounds

The easy interconversion of Fe(II)/Fe(III) has been demonstrated previously (see Experiment 6.7(f)(i), p. 79, and Experiment 8.12(a)(iii), p. 121). The action of sodium hydroxide solution, aqueous ammonia and ammonium thiocyanate solution on aqueous solutions of iron(II) and iron(III) salts has been dealt with in Section 10.5(b)(i)(ii) and (iv) (p. 150). Some complexes of iron(III) have been discussed in Section 10.7 (p. 154).

11.5 Cobalt

Cobalt has oxidation states of +3, which is common in complexes stabilised by ligands such as CN^-, NH_3 and $NH_2CH_2CH_2NH_2$, and +2, which is the most stable oxidation state of simple cobalt compounds.

(a) Cobalt(III) compounds

The best known example of a 'simple' cobalt(III) compound is the fluoride, $Co^{3+}(F^-)_3$. This compound is a strong oxidising agent and reacts with water to form cobalt(II) at the same time liberating oxygen, a behaviour consistent with the following standard redox potential:

$$Co^{3+}(aq) + e^- \rightarrow Co^{2+}(aq) \qquad E^\ominus = +1\cdot82\,V$$

However, this redox potential is reduced considerably in the presence of some ligands, particularly those that contain nitrogen atoms, e.g. NH_3 and $NH_2CH_2CH_2NH_2$, and this accounts for the stability of some cobalt(III) complexes, e.g.

$$[Co(NH_3)_6]^{3+} + e^- \rightarrow [Co(NH_3)_6]^{2+} \qquad E^\ominus = +0\cdot10\,V$$

The following two experiments illustrate the formation of cobalt(III) compounds which exhibit respectively geometrical and optical isomerism.

(i) *Preparation of dichlorobis(1,2-diaminoethane)cobalt(III) chloride*, $[Co(en)_2Cl_2]^+Cl^-$

There are two forms of the compound $[Co(en)_2Cl_2]^+Cl^-$, and these are shown opposite.

Dissolve 16 g of hydrated cobalt(II) chloride in 50 cm³ of distilled water and add, with stirring, a solution of 6·7 cm³ of 1,2-diaminoethane in 50 cm³

trans-isomer (green) cis-isomer (violet)

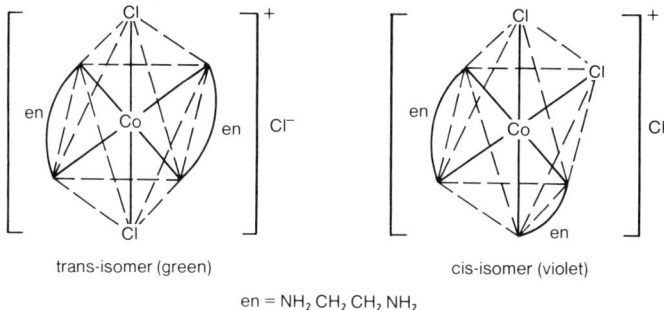

en = NH$_2$CH$_2$CH$_2$NH$_2$

of distilled water to the cool solution of the cobalt salt. Add 10 cm^3 of concentrated hydrochloric acid and draw air through the solution for about 10 hours (or preferably, bubble oxygen from a cylinder through the solution for about 2 hours). Now add 35 cm^3 of concentrated hydrochloric acid to the purple solution and evaporate over a beaker of boiling water until a crust forms on the surface of the solution. Cool the resulting solution in an ice-water mixture and filter off the green crystals using a Buchner funnel. Wash them with ethanol and then with ethoxyethane (diethyl ether). The product is trans-[Co(en)$_2$Cl$_2$]HCl$_2$. Heat the crystals in an oven set to a temperature of 110°C for about 2 hours to obtain trans-[Co(en)$_2$Cl$_2$]$^+$Cl$^-$.

Dissolve some of the final product in distilled water and evaporate the solution over a beaker of boiling water. The solution should turn violet and violet crystals of cis-[Co(en)$_2$Cl$_2$]$^+$Cl$^-$ should begin to crystallise out. Cool in an ice-water mixture and dry the crystals with filter paper.

Although the trans-isomer is more stable than the cis-isomer, the latter is less soluble in water; thus, crystallisation affords a means of isolating the cis-form. Notice that the trans-isomer has a plane of symmetry that is lacking in the cis-isomer, hence the cis-isomer is potentially optically active.

(ii) *Preparation of the optical isomers of tris(1,2-diaminoethane) cobalt(III) iodide monohydrate,* [Co(en)$_3$]$^{3+}$(I$^-$)$_3$.H$_2$O

A cobalt(III) complex containing three 1,2-diaminoethane molecules is potentially optically active and can be resolved into its *d-* and *l-*forms (see below).

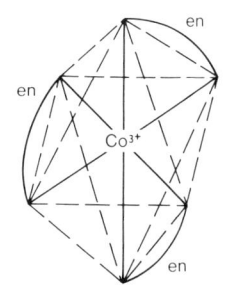

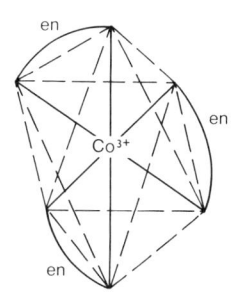

The object of the experiment is to prepare the potentially optically active ion $[Co(en)_3]^{3+}$, to resolve it using barium d-tartrate and finally to prepare the d- and l-forms of $[Co(en)_3]^{3+}(I^-)_3 \cdot H_2O$.

Preparation of the $[Co(en)_3]^{3+}$ ion

Dissolve 11·5 cm³ of 1,2-diaminoethane, $NH_2CH_2CH_2NH_2$, in 25 cm³ of distilled water. After cooling the solution in ice add 8 cm³ of concentrated hydrochloric acid, 14 g of cobalt(II) sulphate, $Co^{2+}SO_4^{2-} \cdot 7H_2O$, in 25 cm³ of distilled water and 2 g of activated charcoal (this catalyses the oxidation of the initial cobalt(II) complex). Draw a current of air through the solution using a water pump for about 4 hours and then adjust the pH of the solution to 7·0–7·5 using approximately 2M hydrochloric acid or 1,2-diaminoethane (check with pH paper). Warm the solution over a beaker of boiling water for about 15 minutes then cool it and filter off the charcoal. Wash the charcoal with about 10 cm³ of distilled water, adding the washings to the filtrate. Keep the filtrate which is a solution of tris(1,2-diaminoethane) cobalt(III) chloride sulphate, $[Co(en)_3]^{3+}Cl^-SO_4^{2-}$.

Preparation of barium d-tartrate

Dissolve 12 g of barium chloride crystals, $Ba^{2+}(Cl^-)_2 \cdot 2H_2O$, and 14 g of sodium potassium tartrate separately in the minimum amounts of hot water at 90°C. Mix the solutions at 90°C cool and filter. Wash the precipitate of barium d-tartrate with a little distilled water and keep it for the next part of the experiment.

Preparation of ($\pm$)tris(1,2-diaminoethane)cobalt(III) chloride d-tartrate

Place the solution of tris(1,2-diaminoethane) chloride sulphate in an evaporating dish and add the barium d-tartrate (including the filter paper). Heat the mixture over a beaker of boiling water for about 30 minutes with constant stirring. Filter off the precipitate of barium sulphate through a fluted paper, washing it with a little hot water and adding the washings to the filtrate. Evaporate the filtrate down to a volume of about 30 cm³ and allow it to stand overnight. Filter off the crystals of (+)tris(1,2-diaminoethane) cobalt(III) chloride d-tartrate. Keep these and also the filtrate, which contains (−)tris(1,2-diaminoethane) cobalt(III) chloride d-tartrate. Wash the crystals with aqueous ethanol and recrystallise from hot water (15 cm³) by cooling in ice. Wash the crystals with aqueous ethanol and then with ethanol.

Preparation of the isomers tris(1,2-diaminoethane)cobalt(III) iodide monohydrate, $[Co(en)_3]^{3+}(I^-)_3 \cdot H_2O$

Dissolve the (+)tris(1,2-diaminoethane) cobalt(III) chloride d-tartrate in 15 cm³ of hot water and add 0·25 cm³ of concentrated ammonia solution followed by 17 g of sodium iodide in 7 cm³ of hot water. Stir the solution. Cool in ice, filter and suck the crystals dry. Wash them with ice-cold sodium iodide solution (3 g in 10 cm³) to remove any tartrate, and then with ethanol and propanone (acetone). The product is (+)tris(1,2-diaminoethane) cobalt(III) iodide monohydrate.

The compound (−)tris(1,2-diaminoethane) cobalt(III) iodide monohydrate is prepared from the filtrate (−)tris(1,2-diaminoethane) cobalt(III) chloride d-tartrate in an exactly similar manner. Finally, dissolve the crude product in 40 cm³ of distilled water at 50°C, filter off undissolved matter and then add 5 g of sodium iodide to the warmed filtrate and allow to crystallise. Filter and wash the crystals with ethanol and propanone (acetone).

$$(\pm)[Co(en)_3]^{3+}Cl^-SO_4^{2-}$$
$$|$$
$$Ba^{2+}d\text{-}C_4H_4O_6^{2-}$$
(barium d-tartrate)

$(+)[Co(en)_3]^{3+}Cl^-d\text{-}C_4H_4O_6^{2-}$ ⟵   ⟶ $(-)[Co(en)_3]^{3+}Cl^-d\text{-}C_4H_4O_6^{2-}$
(crystals)   (filtrate)
| |
Na⁺I⁻   Na⁺I⁻
↓   ↓
$(+)[Co(en)_3]^{3+}(I^-)_3 \cdot H_2O$ $(-)[Co(en)_3]^{3+}(I^-)_3 \cdot H_2O$

If a polarimeter is available, compare the optical rotations of saturated aqueous solutions of the two optical isomers using the sodium D-line.

(b) Cobalt(II) compounds

This is the most stable state of 'simple' cobalt compounds. The action of sodium hydroxide solution, aqueous ammonia, hydrogen sulphide and ammonium thiocyanate solution on aqueous solutions of a cobalt(II) salt has been dealt with in Section 10.5, p. 150. Complexes containing cobalt(II) have been demonstrated in Sections 10.6, p. 152, and 10.7(b)(i)(ii), p. 156.

11.6 Nickel

Although a few compounds are known that contain nickel(III) (e.g. an impure oxide, $Ni_2O_3 \cdot 2H_2O$) and nickel(IV) (e.g. an impure oxide, NiO_2, and a complex fluoride, $(K^+)_2[NiF_6]^{2-}$), the only important oxidation state of nickel is +2.

(a) Nickel(II) compounds

The action of sodium hydroxide solution, aqueous ammonia and hydrogen sulphide on aqueous solutions of a nickel(II) salt has been dealt with in Section 10.5, p. 151. The formation of an EDTA complex with Ni^{2+} ions has been demonstrated in Section 10.7(b)(ii), p. 156. The following experiment describes the determination of the formula of the complex ion formed between EDTA and Ni^{2+} ions by the method of continuous variation.

(i) *Determination of the formula of the* Ni(II)-EDTA *complex ion by colorimetry*

The Ni(II)-EDTA complex ion is blue in aqueous solution, i.e. it absorbs light towards the red end of the electromagnetic spectrum, and its concentration may be measured by determining the extent to which it absorbs this light. The technique is called colorimetry.

The extent to which a substance absorbs monochromatic light is given by the Lambert–Beer Law:

$$I = I_0 e^{-act}$$

where I is the transmitted light intensity, I_0 is the incident light intensity, t is the thickness of the absorbing substance, c is the concentration of absorbing substance and a is a constant. It is more conveniently used in the form:

$$\log(I_0/I) = act/2 \cdot 303 = \varepsilon ct$$

where ε is $a/2 \cdot 303$ and is called the molecular extinction coefficient and $\log(I_0/I)$, called the absorbance, is seen to be directly proportional to the concentration of a particular absorbing substance (for a given thickness of absorbing material).

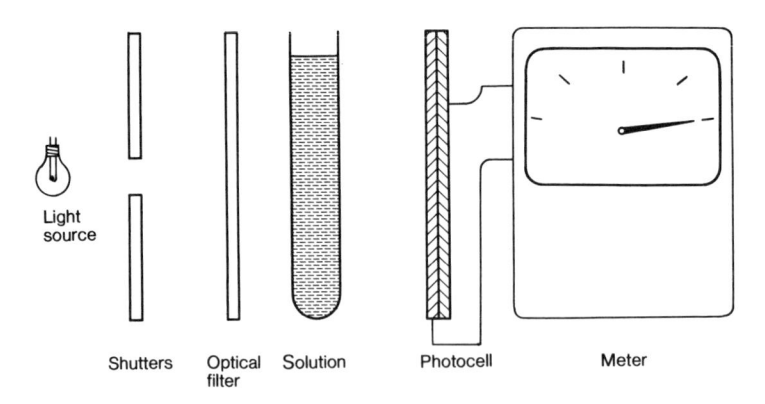

Fig. 11.1
A schematic diagram of a colorimeter.

The colorimeter consists of:

1. A light source.
2. A series of optical filters to isolate a narrow bandwidth. In practice the optical filter selected is the one which allows the absorbing solution to show maximum absorbance.
3. Shutters to control the light intensity. These can be used to set the meter reading to zero absorption with distilled water in the light beam (maximum meter reading).

4. A sample absorbing light in direct proportion to the concentration of the substance being determined (this is an ideal situation never attained in practice).
5. A photocell generating an electric current in direct proportion to the intensity of light transmitted through the sample (again an ideal situation).
6. Output to a meter.

Assuming the Lambert–Beer Law to be applicable to a real situation (it is impossible to select monochromatic light with an optical filter, i.e. light passed through the filter has a finite bandwidth) and condition (5) above to apply, then,

$$\log(I_0/I) = \log(m_0/m) = \varepsilon c t$$

where m_0 and m are the meter readings for the distilled water and the absorbing solution respectively. A plot of $\log(m_0/m)$ against the concentration of the absorbing species should thus ideally be a straight line, and the system under investigation approaches this condition fairly closely.

An Eel Colorimeter Block connected to a 12 V mains transformer is suitable (with the 2 V, 2 W bulb replaced by a 12 V, 2·2 W type). A red Ilford Spectrum Filter (No. 607) is used and the meter (full scale deflection 50 μA) is one made by British Physical Laboratories, Radlett, Herts. The apparatus is shown in Plate 3. Two solutions one 0·05M with respect to Ni^{2+} ions and the other 0·05M with respect to EDTA are required.

Place 11 colorimeter tubes in a rack and make up the following mixtures:

```
 0 cm³ 0·05M Ni²⁺ solution + 10 cm³ 0·05M EDTA solution
 1   ,,                    +  9  ,,
 2   ,,                    +  8  ,,
 3   ,,                    +  7  ,,
 4   ,,                    +  6  ,,
 5   ,,                    +  5  ,,
 6   ,,                    +  4  ,,
 7   ,,                    +  3  ,,
 8   ,,                    +  2  ,,
 9   ,,                    +  1  ,,
10   ,,                    +  0  ,,
```

Choose the red filter No. 607 and then find the meter reading m, for each mixture, adjusting the colorimeter reading by means of the hand-wheel (labelled A in Plate 3) to give $m_0 = 10$ (right hand side of outer black scale), with a tube of distilled water in place before each determination. Before taking a reading, make sure that the colorimeter cover is in place to prevent extraneous light reaching the photo-electric cell. **N.B. In order to avoid damaging the meter, do not switch on the light unless there is a filter in the Colorimeter Block.**

Plot a graph of $\log(m_0/m)$ against the concentration ratios Ni^{2+}/EDTA, i.e. 0/10, 1/9, 2/8 etc., and determine the ratio which gives maximum

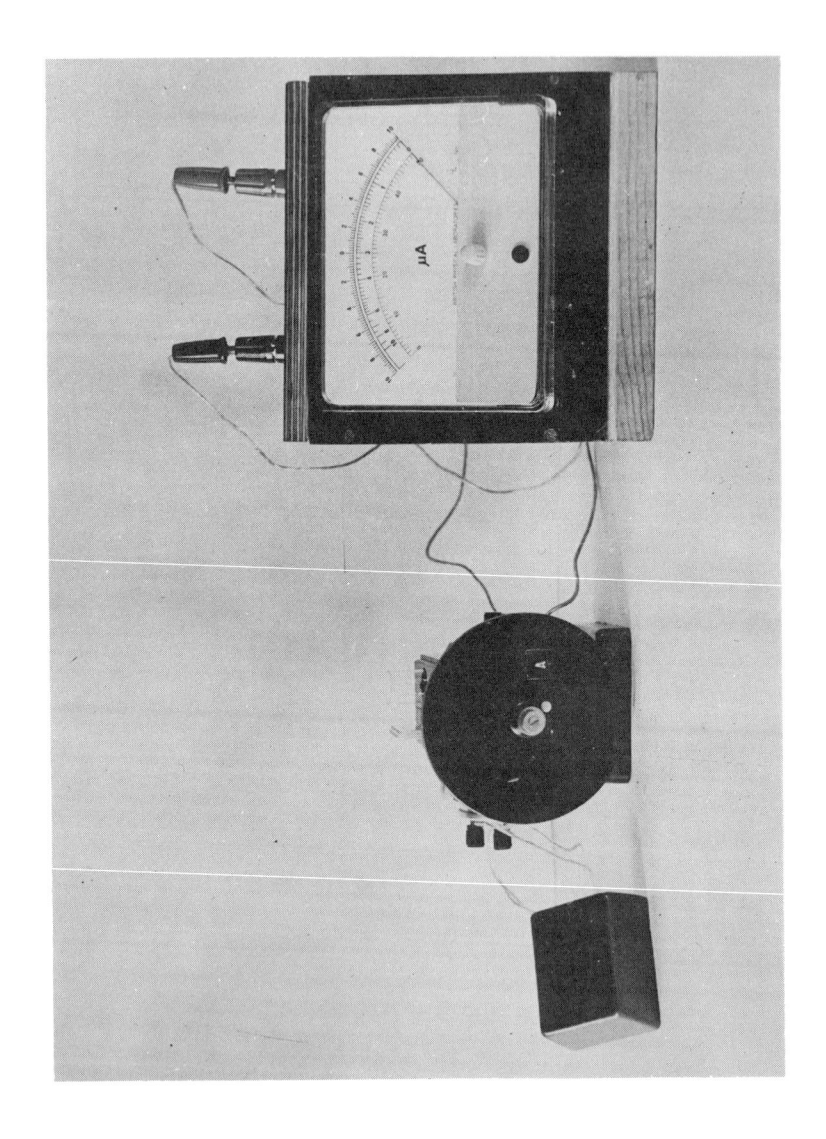

Plate 3
The apparatus used for colorimetric experiments.

absorbance. This ratio is the ratio of Ni^{2+}/EDTA in the complex; what is the formula for the complex ion?

As an alternative the same procedure may be used to determine the formula of the Cu(II)/1,2-diaminoethane complex, using 0·01M copper(II) sulphate solution and 0·01M 1,2-diaminoethane with a green filter. What is the value of n in the complex $[Cu(en)_n]^{2+}$?

(ii) *Formation of bis(dimethylglyoximato)nickel(II)*, $Ni(C_4H_7N_2O_2)_2$

Place about 3 cm³ of nickel sulphate solution in a test-tube and make it just alkaline by the addition of approximately 2M aqueous ammonia (test with indicator paper). Now add an ethanolic solution of dimethylglyoxime and note the formation of a red precipitate of bis(dimethylglyoximato)-nickel(II), warming if necessary. This reaction is quantitative and may be used for the estimation of nickel(II) salts.

bis(dimethylglyoximato)nickel(II)

If dimethylglyoxime is represented as DMG, the equation for its formation can be represented as:

$$2DMG + Ni^{2+} + 2OH^- \rightarrow Ni(DMG)_2 + 2H_2O$$

During the reaction, two protons (one from each dimethylglyoxime molecule) are neutralised by the ammonia solution. The complex, which is thus uncharged, is held in the planar configuration by two hydrogen bonds (shown by dotted lines).

12

GROUP 1B COPPER, SILVER AND GOLD

12.1 Some Physical Data of Group 1B Elements

	Atomic number	Electronic configuration	Standard redox potential/V $M^+ + e^- \rightarrow M$	Standard redox potential/V $M^{n+} + ne^- \rightarrow M$	Density/ g cm^{-3}	M.p./ K	B.p./ K	Atomic radius/ nm	Ionic radius nm M^+
Cu	29	2.8.18.1 ...$3s^2 3p^6 3d^{10} 4s^1$	+0.52	$Cu^{2+} + 2e^- \rightarrow Cu$ +0.34	8.92	1356	2868	0.117	0.096
Ag	47	2.8.18.18.1 ...$4s^2 4p^6 4d^{10} 5s^1$	+0.80		10.5	1233	2485	0.134	0.126
Au	79	2.8.18.32.18.1 ...$5s^2 5p^6 5d^{10} 6s^1$	+1.68	$Au^{3+} + 3e^- \rightarrow Au$ +1.42	19.3	1336	3239	0.134	0.137

12.2 Some General Remarks about Group 1B

The atoms of these three elements have one electron in the outer shell like the atoms of the Group 1A metals but, whereas the atoms of the latter group of elements have eight electrons in the penultimate shell, the atoms of copper, silver and gold have penultimate shells containing eighteen electrons:

Group 1A		Group 1B	
Potassium	2.8.8.1	Copper	2.8.18.1
Rubidium	2.8.18.8.1	Silver	2.8.18.18.1
Caesium	2.8.18.18.8.1	Gold	2.8.18.32.18.1

The fact that the atom of a Group 1B metal has a much larger atomic number, but the same number of electronic shells as its Group 1A counterpart, has important consequences and leads to these two groups of metals having a completely different chemistry. For instance, the nucleus of the copper atom carries a charge of 29+ and has a much greater influence on its surrounding electrons than does the nucleus of the potassium atom with a charge of 19+; consequently the copper atom has a smaller atomic radius and a larger first ionisation energy than the potassium atom. The densities of comparable members of the two groups are widely different

(a) because the Group 1B atoms are heavier and (b) because the atomic radii of Group 1B metals are smaller; the much higher melting points of the Group 1B metals can be attributed (i) to their heavier atoms, and (ii) to the fact that d electrons participate in inter-atomic bonding in addition to the single outer electron. The important differences in properties between Groups 1A and 1B are given in Table 12A.

TABLE 12A   Some differences in properties between Groups 1A and 1B

	K	Rb	Cs	Cu	Ag	Au
Atomic Radius/nm (M^+)	0·203	0·216	0·235	0·117	0·134	0·134
First Ionisation Energy/kJ mol^{-1}	418	403	374	745	737	887
Standard Redox Potential/V	−2·92	−2·99	−3·02	+0·52	+0·80	+1·68
Melting point/K	337	312	302	1356	1233	1336
Density/g cm^{-3}	0·87	1·53	1·90	8·92	10·5	19·3

Copper, silver and gold are known as the 'coinage metals' and are resistant to chemical attack (gold in particular is chemically very unreactive). All three metals exhibit an oxidation state of +1, and in this state their compounds are largely covalent, or at least possess a considerable degree of covalent character, although the principal oxidation states for copper, silver and gold are +2, +1 and +3 respectively. In the higher oxidation states the Group 1B metals utilise their d electrons in addition to their single s electron, thus their ions contain an incomplete d level and are typically transitional. Unlike the Group 1A metals, copper, silver and gold form numerous complexes in all their oxidation states.

Since there are no smooth gradations in properties along the series, copper, silver and gold, the chemistry of copper and silver are considered separately; no experimental work on gold is included in this chapter.

COPPER

12.3 Extraction of Copper

Copper is principally extracted from copper pyrites, $CuFeS_2$, copper glance, Cu_2S, and cuprite, Cu_2O; it is also mined as the free element. The extraction of the metal from cuprite involves purification followed by controlled heating in air:

$$Cu_2S + O_2 \rightarrow 2Cu + SO_2$$

It is refined by electrolytic methods.

12.4 Some Reactions of Copper

At about 570K copper is attacked by air or oxygen, and a black coating of copper(II) oxide forms on its surface; at a temperature of about 1270K copper(I) oxide is formed instead. Copper is also attacked by sulphur vapour, with the formation of copper(I) sulphide and by chlorine which forms copper(II) chloride.

(a) Action of sulphur vapour on copper

Place about 1·0 g of powdered roll sulphur in a test-tube and place a small plug of 'Rocksil' wool loosely above it. On top of the 'Rocksil' wool place about 1·5 g of copper filings. Heat the sulphur until it begins to vaporise and note any reaction with the copper. Record your observations.

(b) Action of acids on copper

(i) Action of nitric acid on the metal

Place about 0·5 g of copper filings in a test-tube and then add about 3 cm³ of approximately 2M nitric acid. Is there any sign of reaction? Warm if necessary and record your results.

Place about 0·5 g of copper filings in a crucible and then **cautiously** add about 2 cm³ of concentrated nitric acid. **This reaction should be carried out in a fume cupboard.** Record your observations.

Consult a testbook of inorganic chemistry for an explanation of the main reactions involved in these two experiments.

(ii) Action of hydrochloric acid on the metal

Place about 0·5 g of copper filings in a test-tube and then add about 3 cm³ of approximately 2M hydrochloric acid. Warm if necessary. Are there any signs of reaction? To the warm mixture add a few drops of 20 vol. hydrogen peroxide solution and note the result. Can you give any explanation of the action of hydrogen peroxide in the above reaction?

Repeat the above experiment using concentrated hydrochloric acid.

COMPOUNDS OF COPPER(II)

This is the most common oxidation state of copper and in aqueous solution copper(II) salts are blue, the colour being due to the presence of $[Cu(H_2O)_6]^{2+}$ ions.

12.5 Some Complex Ions Containing Copper(II)

Copper in the oxidation state +2 forms many complex ions which exhibit a variety of colours. In the following experiments the major complex ion species present are $[CuCl_4]^{2-}$, $[CuBr_4]^{2-}$, $[Cu(H_2O)_6]^{2+}$ and $[Cu(NH_3)_4]^{2+}$.

(a) Some visual observations involving copper(II) complexes

Place about 1 g of copper(II) chloride dihydrate, $Cu^{2+}(Cl^-)_2 \cdot 2H_2O$ in a test-tube and then add to it about $2\,cm^3$ of concentrated hydrochloric acid. Warm to obtain a solution and, if some solid still remains undissolved, add the minimum of distilled water to obtain a solution. What colour is the solution which contains the $[CuCl_4]^{2-}$ complex ion?

To the above solution add, in portions, some solid potassium bromide (a source of Br^- ions) until some remains undissolved after shaking. What now is the colour of the solution that contains $[CuBr_4]^{2-}$ ions? Dilute this solution gradually with water and notice any colour changes, stopping the addition of water when a light blue solution is obtained (due to the presence of the $[Cu(H_2O)_6]^{2+}$ ion). Now add approximately 2M aqueous ammonia continuing the addition of the aqueous ammonia until the initial precipitate has dissolved. Note the final colour of the solution which is due to the presence of the $[Cu(NH_3)_4]^{2+}$ ion.

Write equations for the successive formation of the above complex ions. Do you think the above changes are reversible? Find out by experiment if they are. When bromide ions are added to the solution containing $[CuCl_4]^{2-}$, what complex ions would you expect to be formed before the major ion $[CuBr_4]^{2-}$ is formed?

12.6 Copper(II) Hydroxide, $Cu(OH)_2$, and Copper(II) Oxide, CuO

(a) Formation of copper(II) hydroxide and some of its reactions

Place about $3\,cm^3$ of a solution of copper(II) sulphate in a test-tube and add about the same volume of approximately 2M sodium hydroxide solution. Note the colour and general appearance of the precipitate (it can be filtered and dried at $100°C$ to the composition $Cu(OH)_2$, if an excess of alkali is used). Divide the mixture into two equal portions. To one portion add approximately 2M aqueous ammonia until no further change occurs, and observe what happens. Explain, in terms of equations, what happens in the presence of aqueous ammonia, given that

$$\underset{\text{solid}}{Cu(OH)_2} \rightleftharpoons \underset{\text{in solution}}{Cu^{2+} + 2OH^-}$$

Dilute the other portion of the mixture containing copper hydroxide with an equal volume of distilled water and then heat. Note any colour changes and suggest what the final residue is? See if this residue is also soluble in aqueous ammonia.

(b) Formation of copper(II) oxide

Place about $0.5\,g$ of copper(II) carbonate in an ignition tube and heat gently until no further change occurs. See if the residue of copper(II) oxide is soluble in approximately 2M aqueous ammonia.

Repeat the experiment using about 0·5 g of copper(II) nitrate taking care not to breath any of the brown gas, nitrogen dioxide. What other gas is evolved? Write an equation for this thermal decomposition reaction.

12.7 Copper(II) Sulphide, CuS

(a) Formation of copper(II) sulphide

Place about 3 cm³ of copper(II) sulphate solution in a test-tube and pass into it a stream of hydrogen sulphide. Note the colour of the copper(II) sulphide precipitate. Find out whether the precipitate is soluble in hydrochloric acid (try both dilute and concentrated acid). Explain your result (zinc sulphide, for example, is soluble in dilute hydrochloric acid).

12.8 Copper(II) Carbonate, $aCuCO_3 . bCu(OH)_2$

The pure substance is unknown, any method of preparation, e.g. the addition of either sodium carbonate or sodium hydrogen carbonate solutions to a solution of a copper(II) salt, resulting in the formation of a green basic carbonate. Can you explain why?

(a) Solubility of copper(II) carbonate in aqueous ammonia

Place about 0·2 g of copper(II) carbonate in a test-tube and add about 5 cm³ of approximately 2M aqueous ammonia. Shake and filter if necessary to obtain a clear solution. Is copper(II) carbonate appreciably soluble in aqueous ammonia?

Now pass a stream of hydrogen sulphide through the solution and note what happens. Attempt to explain all the above reactions in terms of equations.

12.9 Oxidising Action of Copper(II) Salts

(a) Action of a solution of copper(II) sulphate on a solution of potassium iodide

Place 20 drops of 0·1M copper(II) sulphate solution in a test-tube using a dropping tube. Now add about 10 drops of an approximately 2M solution of potassium iodide using the same clean dropping tube (this represents a considerable excess). Note what happens, one of the products being iodine. Now add 0·1M sodium thiosulphate solution, drop by drop, using the same clean dropping tube, shaking the test-tube after the addition of each drop. After adding about 15 drops of the sodium thiosulphate solution, add a few drops of fresh starch solution and continue adding the sodium thiosulphate solution, drop by drop, until the dark colour of the starch-iodine is discharged. Note the number of drops added. What do you think is the white solid that remains?

Given that iodine reacts with sodium thiosulphate solution according to the equation:

$$I_2 + 2S_2O_3^{2-} \rightarrow 2I^- + S_4O_6^{2-}$$

calculate the number of moles of copper(II) sulphate (to the nearest whole number) that produce 1 mole of iodine on reaction with potassium iodide solution; hence try to construct a balanced equation for the reaction between solutions of copper(II) sulphate and potassium iodide.

(b) Addition of a solution of copper(II) sulphate to a solution of potassium bromide

Place about 3 cm³ of a solution of potassium bromide in a test-tube and then add about the same volume of copper(II) sulphate solution. Is the result similar to that in Experiment 12.9(a)? If not, can you suggest a reason why not?

COMPOUNDS OF COPPER(I)

In aqueous solution the hydrated copper(I) ion is unstable and disproportionates into the copper(II) ion and copper, i.e. undergoes self oxidation-reduction; this is indicated by the standard redox potentials for the systems $Cu^+(aq)/Cu$ and $Cu^{2+}(aq)/Cu^+(aq)$ which are given below:

$$Cu^+(aq) + e^- \rightarrow Cu \qquad E^\ominus = +0.52 \text{ V}$$

$$Cu^{2+}(aq) + e^- \rightarrow Cu^+(aq) \qquad E^\ominus = +0.16 \text{ V}$$

or
$$2Cu^+(aq) \rightarrow Cu + Cu^{2+}(aq) \qquad E^\ominus_{total} = +0.36 \text{ V}$$

The positive e.m.f. of the above cell reaction implies that hydrated copper(I) ions are unstable in solution with respect to copper and hydrated copper(II) ions. The value of the equilibrium constant for this disproportionation reaction has been estimated to be in the order of $10^6 \text{ dm}^3 \text{ mol}^{-1}$ at 298K.

$$2Cu^+(aq) \rightleftharpoons Cu + Cu^{2+}(aq)$$

$$K = [Cu^{2+}(aq)]/[Cu^+(aq)]^2 = 10^6 \text{ dm}^3 \text{ mol}^{-1}$$

i.e. the concentration of the hydrated Cu^+ ions in solution is extremely small. The equilibrium can be shifted to the left by adding anions which precipitate out an insoluble copper(I) compound, for example, I^- ions precipitate insoluble CuI (see Experiment 12.9(a), p. 182), or by adding a substance which forms a more stable complex ion with Cu^+ than with Cu^{2+} (see Experiment 12.11, p. 184).

Although the chemistry of copper(I) is largely that of its water-insoluble compounds and of its stable complexes, other copper(I) compounds are perfectly stable in the absence of moisture, e.g. copper(I) sulphate, Cu_2SO_4.

12.10 Copper(I) Oxide, Cu_2O

(a) Formation of copper(I) oxide

Copper(I) oxide is obtained as a red solid by the reduction of an alkaline solution of copper(II) sulphate. Since the addition of alkali to a solution of a copper(II) salt would result in the precipitation of copper(II) hydroxide, the copper(II) ions are complexed with tartrate ions; under these conditions the Cu^{2+} ions are present in such low concentration that the solubility product of copper(II) hydroxide is not exceeded.

To about 2 cm^3 of an approximately 0·3M solution of copper(II) sulphate add the same volume of an alkaline solution of sodium potassium tartrate (approximately 1·3M with respect to sodium potassium tartrate and 4M with respect to sodium hydroxide). A deep blue solution should be obtained, called Fehling's solution, which contains a copper(II) tartrate complex. To this solution add about 0·3 g of glucose (the reducing agent) and warm for several minutes. Note what you observe.

(b) Disproportionation of copper(I) oxide in the presence of dilute sulphuric acid

Place about 0·2 g of copper(I) oxide in a test-tube and then add to it about 3 cm^3 of approximately M sulphuric acid. Warm for several minutes and then allow the solid to settle. Complete and balance the equation:

$$Cu_2O + H^+ \rightarrow$$

12.11 Copper(I) Chloride, CuCl

(a) Formation of copper(I) chloride

Place about 1·5 g of copper turnings and 1 g of copper(II) chloride dihydrate, $Cu^{2+}(Cl^-)_2 \cdot 2H_2O$, in a boiling-tube and then add to the mixture about 4 cm^3 of concentrated hydrochloric acid. Heat the boiling-tube in a beaker of boiling water for about 10 minutes, preferably in a fume cupboard, and then pour the mixture into about 75 cm^3 of cold distilled water in a small beaker. Note the formation of a white precipitate of copper(I) chloride. Filter off the solid product using a Buchner filter and scrape most of the solid into a test-tube for the following experiment. Leave some of the solid exposed to the air and observe and explain what happens.

(b) Complexes of copper(I)

Copper(I) chloride is soluble in water in the presence of entities such as Cl^-, NH_3, $S_2O_3^{2-}$ and Br^- with which it forms complex ions, e.g. $[Cu(NH_3)_2]^+$.

Add about 5 cm^3 of distilled water to the copper(I) chloride prepared in Experiment 12.11(a) and shake to obtain a suspension. Divide this suspension into four equal portions. To one portion add concentrated hydrochloric acid until no further change occurs; to the second portion add approximately 2M aqueous ammonia until a solution is obtained; to the third add solid potassium bromide in portions with shaking, and to the fourth add sodium thiosulphate solution until the solid dissolves.

The complex ion $[Cu(NH_3)_2]^+$ is colourless, but your solution will probably be blue. Can you explain why?

12.12 Copper(I) Iodide, CuI

Copper(I) iodide is precipitated as a white solid when a solution of potassium iodide is added to a solution of copper(II) sulphate (see Experiment 12.9(a)):

$$2Cu^{2+} + 4I^- \rightarrow 2CuI + I_2$$

Copper(II) iodide has never been prepared. Can you explain why not?

SILVER

12.13 Extraction of Silver

Silver occurs as silver sulphide, Ag_2S, and silver chloride, $AgCl$; it also occurs as the free element. Significant amounts of silver are also obtained during the extraction of lead from its ores and during the electrolytic refining of copper.

One method of extraction involves the treatment of the ore with an aerated solution of sodium cyanide. The silver is taken into solution as the $[Ag(CN)_2]^-$ complex from which silver can be displaced by treatment with zinc (see an inorganic textbook for details). A similar technique is used for extracting gold.

12.14 Some Reactions of Silver

The metal is resistant to attack by the air and moisture, although the familiar black stain of silver sulphide results from exposure of the metal to hydrogen sulphide. Steam and dilute non-oxidising acids have no effect on the metal but it is attacked by hot concentrated sulphuric acid and cold dilute nitric acid with the formation of silver(I) ions, Ag^+:

$$2Ag + 2H_2SO_4 \rightarrow Ag_2SO_4 + SO_2 + 2H_2O$$

$$3Ag + 4HNO_3 \rightarrow 3AgNO_3 + 2H_2O + NO$$

Concentrated nitric acid produces mainly nitrogen dioxide and not nitrogen oxide.

12.15 The Oxidation States of Silver

The important oxidation state of silver is $+1$ although compounds of silver(II) (e.g. AgF_2) and silver(III) (e.g. $K^+[AgF_4]^-$), are known.

Compounds containing silver(III) or silver(II) are powerful oxidising agents. The instability of $Ag^{2+}(aq)$ with respect to $Ag^+(aq)$ is indicated by the standard redox potentials for the systems $Ag^{2+}(aq)/Ag^+(aq)$ and $Ag^+(aq)/Ag$ which are given below:

$$Ag^{2+}(aq) + e^- \rightarrow Ag^+(aq) \qquad E^\ominus = +1\cdot98 \text{ V}$$
$$Ag^+(aq) + e^- \rightarrow Ag \qquad E^\ominus = +0\cdot80 \text{ V}$$

or $\qquad Ag^{2+}(aq) + Ag \rightarrow 2Ag^+(aq) \qquad E^\ominus_{total} = +1\cdot18 \text{ V}$

The positive e.m.f. of the above cell reaction implies that $Ag^{2+}(aq)$ ions are capable of oxidising silver to $Ag^+(aq)$ ions. Notice that the disproportionation of $Ag^+(aq)$ into $Ag^{2+}(aq)$ and Ag cannot take place since the e.m.f. of this cell reaction is negative (c.f. the disproportionation of $Cu^+(aq)$ into $Cu^{2+}(aq)$ and Cu (p. 183)).

COMPOUNDS OF SILVER(I)

Since the compounds of silver of any importance are those that contain silver(I), it is usual to refer to silver(I) compounds simply as silver compounds.

12.16 Silver Oxide, Ag_2O

Silver hydroxide does not exist and the addition of sodium hydroxide solution to a solution of a silver salt results in the precipitation of silver oxide.

(a) Formation of silver oxide

To about 2 cm^3 of a solution of silver nitrate add about the same volume of a solution of approximately 2M sodium hydroxide and note the result. Keep the mixture for the next experiment.

(b) Solubility of silver oxide in aqueous ammonia

Gradually add approximately 2M aqueous ammonia to the mixture obtained in (a) above and go on adding the ammonia solution until no further change occurs. Can you explain why the silver oxide should be soluble in the aqueous ammonia?

A solution of silver oxide in aqueous ammonia is used to distinguish aldehydes from ketones. Aldehydes, but not ketones, reduce this mixture and, under controlled conditions, a silver mirror may be obtained.

12.17 Silver Halides, AgX

Silver fluoride is a white, water-soluble solid whereas the other halides of silver are only very sparingly soluble in water.

(a) Precipitation of silver chloride, silver bromide and silver iodide

Place about 1 cm³ of sodium chloride, potassium bromide and potassium iodide solutions respectively in three separate test-tubes and then add about the same volume of silver nitrate solution to each. Compare the colours of the silver halide precipitates.

To each tube now add some approximately 2M aqueous ammonia and after shaking compare the solubilities of these three silver halides in the ammonia solution.

(b) Comparison of the solubilities of silver chloride, silver bromide and silver iodide

To about 2 cm³ of a solution of sodium chloride add about the same volume of silver nitrate solution. Now add gradually with stirring some approximately 2M aqueous ammonia solution until the precipitate of silver chloride has dissolved. To this solution add some potassium bromide solution; what happens and why? Now add some sodium thiosulphate solution with stirring until the precipitate of silver bromide has dissolved. To this clear solution now add some potassium iodide solution; what happens and why?

Can you list the silver halides in decreasing order of solubility in water?

Given that aqueous ammonia and sodium thiosulphate solution react with silver ions according to the equations given below, can you say which of the complex ions $[Ag(NH_3)_2]^+$ or $[Ag(S_2O_3)_2]^{3-}$ is the more stable with respect to dissociation into silver ions and either NH_3 or $S_2O_3^{2-}$?

$$Ag^+ + 2NH_3 \rightleftharpoons [Ag(NH_3)_2]^+$$
$$Ag^+ + 2S_2O_3^{2-} \rightleftharpoons [Ag(S_2O_3)_2]^{3-}$$

12.18 Silver Sulphide, Ag₂S

Silver sulphide is exceedingly insoluble in water and can be precipitated from solutions containing only a minute concentration of silver ions, e.g., from solutions in which silver ions are bound in the form of stable complexes.

(a) Precipitation of silver sulphide from a silver complex

To about 2 cm³ of a solution of silver nitrate add sodium chloride solution until all the silver chloride has precipitated. Now add sodium thiosulphate solution until the precipitate of silver chloride has dissolved and then pass hydrogen sulphide through the solution. Observe what happens. Can you explain your observations in terms of the equilibrium given below?

$$Ag^+ + 2S_2O_3^{2-} \rightleftharpoons [Ag(S_2O_3)_2]^{3-}$$

12.19 Silver Chromate, Ag_2CrO_4

(a) Formation of silver chromate and reaction with sodium chloride solution

Place about 2 cm³ of potassium chromate solution in a test-tube and add silver nitrate solution to precipitate silver chromate (a brick-red solid). Now add sodium chloride solution to this mixture, until no further change occurs. Note that the red precipitate of silver chromate is replaced by a white precipitate of silver chloride, since the solubility product of silver chloride is smaller than that of silver chromate.

(b) Use of potassium chromate as an indicator in titrations of neutral chloride solutions with silver nitrate solution

The fact that silver chromate has a higher solubility product than silver chloride is used in the estimation of chloride solutions with silver nitrate solution.

Place 20 drops of 0·1M potassium chloride solution in a test-tube and add 2 drops of approximately 0·25M potassium chromate solution using the same clean dropping tube. Wash the dropping tube with distilled water and then with 0·1M silver nitrate solution, and then add the 0·1M silver nitrate solution, drop by drop, until you notice in the mixture, after shaking, the first appearance of a reddish tinge. Determine the number of moles of silver nitrate solution, to the nearest whole number, that react with one mole of potassium chloride solution.

Silver nitrate solution reacts with potassium chloride solution according to the stoichiometric equation:

$$Ag^+ + Cl^- \rightarrow AgCl$$

Are your results sufficiently close to this to indicate that silver chloride precipitates virtually completely before any precipitate of silver chromate forms?

12.20 Summary

(a) Copper occurs in a number of ores, its extraction from cuprite, Cu_2S, involving controlled heating in air. It is purified by electrolytic methods.

(b) On heating in air, copper suffers surface attack by oxygen; it is also attacked by sulphur vapour. Dilute hydrochloric acid has no effect on the metal, but the metal readily reacts with both dilute and concentrated nitric acid.

(c) Aqueous solutions containing copper(II) compounds are blue, owing to the presence of $[Cu(H_2O)_6]^{2+}$ ions. Addition of aqueous ammonia gives the deep blue $[Cu(NH_3)_4]^{2+}$ ion, and the addition of chloride

ions gives the green $[CuCl_4]^{2-}$ ion. These changes are easily reversed.
(d) Copper(II) hydroxide is precipitated as a light blue solid when an aqueous solution of a copper(II) salt is treated with sodium hydroxide solution. It is soluble in aqueous ammonia and is decomposed by heating to give copper(II) oxide and water.
(e) Copper(II) oxide may be obtained by the thermal decomposition of copper(II) carbonate (a basic salt) or copper(II) nitrate.
(f) Aqueous solutions of copper(II) salts oxidise iodide ions to iodine. A white precipitate of copper(I) iodide is formed as well. Aqueous copper(II) ions are not sufficiently strong oxidising agents to convert bromide ions to bromine.
(g) The chemistry of copper(I) compounds is largely that of its water-insoluble compounds, e.g. copper(I) oxide, and of its stable complexes, e.g. $[Cu(NH_3)_2]^+$.
(h) Copper(I) oxide disproportionates into copper(II) and metallic copper when treated with dilute sulphuric acid.
(i) Copper(I) chloride is a white solid which is slowly oxidised by air to copper(II) chloride. It is soluble in aqueous ammonia, and in aqueous solutions containing chloride ions, thiosulphate ions or bromide ions, with the formation of complex ions. Copper(I) iodide is stable in air, but copper(II) iodide does not exist.
(j) Silver occurs as silver sulphide, Ag_2S, and silver chloride, $AgCl$. Its extraction involves the formation of the $[Ag(CN)_2]^-$ complex ion from which silver is precipitated by the addition of zinc.
(k) Silver is resistant to attack by air and dilute non-oxidising acids, but both dilute and concentrated nitric acid attack it.
(l) The chemistry of silver is almost exclusively that of silver(I) compounds. The oxide, Ag_2O, may be precipitated by the addition of sodium hydroxide solution to a solution of silver salt. It dissolves in aqueous ammonia forming the $[Ag(NH_3)_2]^+$ ion.
(m) Silver fluoride is soluble in water whereas the other silver halides are only very sparingly soluble.
(n) Silver ions form complex ions with aqueous ammonia and with a solution containing thiosulphate ions. The latter has the formula $[Ag(S_2O_3)_2]^{3-}$ and its formation is the basis of 'fixing' in photography.
(o) Neutral chloride solutions may be estimated by the addition of standard silver nitrate solution, which precipitates silver chloride. The indicator used is potassium chromate solution, which forms a red precipitate of silver chromate, Ag_2CrO_4, once all the chloride ions have been precipitated as silver chloride.

13

GROUP 2B ZINC, CADMIUM AND MERCURY

13.1 Some Physical Data of Group 2B Elements

	Atomic number	Electronic configuration	Standard redox potential/V $M^{2+}+2e^-\to M$	Density/ $g\,cm^{-3}$	M.p./ K	B.p./ K	Atomic radius/nm	Ionic radius/nm M^{2+}
Zn	30	2.8.18.2 $...3s^23p^63d^{10}4s^2$	−0·76	7·1	692	1180	0·125	0·074
Cd	48	2.8.18.18.2 $...4s^24p^64d^{10}5s^2$	−0·40	8·6	594	1041	0·141	0·097
Hg	80	2.8.18.32.18.2 $...5s^25p^65d^{10}6s^2$	+0·85	13·6	234	630	0·144	0·110

13.2 Some General Remarks about Group 2B

The atoms of the Group 2B elements have two electrons in the outer shell, like the atoms of the Group 2A elements; but whereas the atoms of the latter group of metals have eight electrons in the penultimate shell those of zinc, cadmium and mercury have penultimate shells containing eighteen electrons:

Group 2A		Group 2B	
Calcium	2.8.8.2	Zinc	2.8.18.2
Strontium	2.8.18.8.2	Cadmium	2.8.18.18.2
Barium	2.8.18.18.8.2	Mercury	2.8.18.32.18.2

There is little in common between comparable members of Groups 2A and 2B.

There is a marked difference in physical properties between Group 2B metals and those of Group 1B (copper, silver and gold); for example, the melting points, boiling points and thermal and electrical conductivities of zinc, cadmium and mercury are very much lower than those of copper, silver and gold. These differences are surprising in view of the fact that both groups of metals are assigned completely filled d levels; clearly the d electrons of the Group 2B metals cannot participate in inter-atomic bonding to any great extent, neither are they readily available for thermal and electrical conductivity, although the precise reason why this should be so is not fully understood. The fact that mercury is a liquid is attributed to the inert nature of the two 6s electrons (the inert pair effect); once again, this effect is not fully understood, but it is a feature of the chemistry of the heavier elements.

Zinc, cadmium and mercury are non-transition metals, i.e. they do not exhibit oxidation states in which d electrons are involved; however, like transition metals they form a wide range of complexes, particularly with ligands such as ammonia, cyanide ions and halide ions. In general, the mercury complexes are the most stable and the zinc complexes the least stable, although this trend is reversed when the bonded atom is oxygen; for example, zinc oxide is amphoteric and forms the zincate anion, $[Zn(OH)_4]^{2-}$, whereas neither cadmium oxide nor mercury(II) oxide is amphoteric. All three metals exhibit a principal oxidation state of $+2$ and, in addition, mercury forms the unique mercury(I) ion, Hg_2^{2+}, in which two mercury atoms are covalently bonded. Mercury in particular displays a marked tendency to form covalent bonds, which accounts for the large number of organometallic compounds formed by this element.

Zinc and cadmium are sufficiently similar to be treated comparatively; there are few points of similarity between cadmium and mercury, and the latter is treated separately.

ZINC AND CADMIUM

13.3 Extraction of Zinc and Cadmium

The principal source of zinc is the sulphide zinc blende, ZnS, although the metal is also extracted from calamine, $ZnCO_3$. There are no workable sources of cadmium and this metal is obtained as a by-product of the zinc industry.

The extraction of zinc from the sulphide ore involves purification, roasting in air and subsequent reduction of zinc oxide to the metal:

$$2ZnS + 3O_2 \rightarrow 2ZnO + 2SO_2$$
$$ZnO + C \rightarrow Zn + CO$$

Cadmium is obtained during the distillation of crude zinc.

13.4 Some Reactions of Zinc and Cadmium

Both zinc and cadmium are fairly reactive metals, and when exposed to moist air for any length of time, a protective layer is formed on their surfaces; this is the oxide initially, but over a period of time the basic carbonate is formed. Non-metals such as oxygen, sulphur and the halogens combine directly with the two metals on heating.

Action of acids on zinc and cadmium

(i) *Action of dilute nitric acid on zinc and cadmium*

Place about 0·5 g of zinc (granulated or foil) in a test-tube and then add about 3 cm^3 of approximately 2M nitric acid. Is there any sign of reaction? Warm if necessary and record your results.

Repeat the experiment with cadmium if any is available.

(ii) *Action of hydrochloric acid on zinc and cadmium*

Place a small piece of pure zinc foil (about 1 cm × 4 cm) in a test-tube and cover it with approximately 2M hydrochloric acid. Is there much evidence of hydrogen evolution? Now touch the surface of the zinc with a piece of copper wire (which by itself does not react with dilute hydrochloric acid to evolve hydrogen). From which metal surface does gas evolution occur? Consult a textbook for an explanation of this phenomenon.

Place about 0·5 g of cadmium in a test-tube and add about 3 cm^3 of approximately 2M hydrochloric acid. Is there any sign of gas evolution? Warm if necessary. Decant the dilute hydrochloric acid and replace it with concentrated hydrochloric acid. Warm if necessary to obtain hydrogen. Write an equation for the action of hydrochloric acid on cadmium.

13.5 Zinc Oxide, ZnO, and Cadmium Oxide, CdO

(a) Formation of zinc oxide and cadmium oxide

Place about 0·5 g of zinc carbonate in a test-tube and heat gently until carbon dioxide evolution ceases. Note the colour of the resulting zinc oxide when it is hot and when it has cooled down. Keep the product for the next experiment.

Repeat the above experiment using cadmium carbonate.

(b) Action of dilute sulphuric acid on zinc oxide and cadmium oxide

Place about half the zinc oxide in a test-tube and add approximately M sulphuric acid. Warm and note whether it reacts to give a clear solution.

Repeat with half the cadmium oxide in a similar manner.

(c) Action of sodium hydroxide solution on zinc oxide and cadmium oxide

Place the other half of the zinc oxide in a test-tube and add approximately 2M sodium hydroxide solution. Warm and note whether it reacts to give a clear solution.

Repeat with the other half of the cadmium oxide in a similar manner.

From the results of Experiments 13.5(b) and (c) can you classify zinc and cadmium oxides as either basic or amphoteric?

13.6 Zinc Hydroxide, Zn(OH)$_2$, and Cadmium Hydroxide, Cd(OH)$_2$

(a) Formation of zinc hydroxide and cadmium hydroxide

Place about 2 cm³ of zinc sulphate solution in a test-tube and add approximately 2M sodium hydroxide solution, drop by drop, until a precipitate forms.

Repeat the experiment, but this time replace the zinc sulphate solution with a solution of cadmium sulphate.

(b) Action of dilute sulphuric acid, sodium hydroxide solution and aqueous ammonia on zinc hydroxide and cadmium hydroxide

Prepare some zinc hydroxide as above and treat separate portions with approximately M sulphuric acid, approximately 2M sodium hydroxide solution and approximately 2M aqueous ammonia. Stir and note whether the zinc hydroxide dissolves in each case.

Repeat the experiment, but replace the zinc hydroxide with cadmium hydroxide. Does the cadmium hydroxide dissolve in all three cases?

Would you classify zinc hydroxide and cadmium hydroxide as basic or amphoteric? Both zinc hydroxide and cadmium hydroxide react with aqueous ammonia to give ammine complexes with formulae:

$$[Zn(NH_3)_4]^{2+} \quad \text{and} \quad [Cd(NH_3)_4]^{2+}$$

13.7 Zinc Sulphide, ZnS, and Cadmium Sulphide, CdS

Formation of zinc sulphide and cadmium sulphide and the effect of hydrochloric acid on them

Place about 3 cm³ of zinc sulphate solution in a test-tube and pass into the solution some hydrogen sulphide. Note the colour of the zinc sulphide precipitate. Add some approximately 2M hydrochloric acid and note whether the zinc sulphide dissolves.

Repeat the experiment with about 3 cm³ of cadmium sulphate solution and note the colour of the cadmium sulphide precipitate. See if the cadmium sulphide dissolves in approximately 2M hydrochloric acid. If it seems to have no effect, try some concentrated hydrochloric acid.

Hydrogen sulphide in aqueous solution ionises as follows:

$$H_2O + H_2S \rightleftharpoons H_3O^+ + HS^-$$
$$H_2O + HS^- \rightleftharpoons H_3O^+ + S^{2-}$$

What will be the effect on the sulphide ion concentration of added hydrochloric acid (H$_3$O$^+$)? In view of what you have observed above, can you say whether zinc sulphide has a higher or lower solubility product than cadmium sulphide?

13.8 Zinc Carbonate, $ZnCO_3$, and Cadmium Carbonate, $CdCO_3$

The pure carbonates can be precipitated by the addition of sodium hydrogen carbonate solution to a solution of a zinc and a cadmium salt. If sodium carbonate solution is used in place of sodium hydrogen carbonate, basic carbonates are precipitated. Can you explain these differences?

13.9 Zinc Nitrate, $Zn(NO_3)_2$, and Cadmium Nitrate, $Cd(NO_3)_2$

Note that both nitrates are very deliquescent.

Action of heat on zinc nitrate and cadmium nitrate

Do these experiments in a fume cupboard. Place about 0·5 g of zinc nitrate and cadmium nitrate in separate test-tubes and heat until no further change occurs. What gases are evolved and what are the solid residues in each case?

MERCURY

13.10 Extraction of Mercury

Mercury is extracted from the mineral cinnabar, HgS, the deposits in Spain and Italy accounting for about three-quarters of the world's supply of the metal. The sulphide is heated in air, the mercury vapour which is evolved being condensed to the liquid metal:

$$HgS + O_2 \rightarrow Hg + SO_2$$

13.11 Some Reactions of Mercury

Pure mercury is not attacked by air at ordinary temperatures; mercury(II) oxide is slowly formed in the region of about 620K but at slightly higher temperatures it decomposes into mercury and oxygen; other non-metals that combine directly with mercury include the halogens and sulphur.

In view of its high positive electrode potential, mercury is not attacked by dilute non-oxidising acids (contrast the behaviour of zinc and cadmium). Nitric acid attacks the metal with the liberation of oxides of nitrogen; excess of concentrated nitric acid tends to give nitrogen dioxide and mercury(II) nitrate, but with dilution, and in the presence of excess mercury, nitrogen oxide and mercury(I) nitrate tend to predominate.

COMPOUNDS OF MERCURY

Mercury exhibits oxidation states of $+2$ and $+1$; in the latter, two mercury atoms are united by a single covalent bond as in the $[Hg-Hg]^{2+}$ ion. Mercury compounds are more covalent than those of zinc and cadmium and, indeed, covalency is the rule rather than the exception. Mercury(II) compounds form stable complexes with a variety of ligands, and they are generally many orders of magnitude more stable than the corresponding complexes of zinc and cadmium.

In view of the toxic properties of mercury vapour and mercury compounds the following experiments are best done as demonstrations.

COMPOUNDS OF MERCURY(II)

13.12 Mercury(II) Oxide, HgO

(a) Formation of mercury(II) oxide

Place about $2 cm^3$ of a solution of mercury(II) chloride in a test-tube and add to it about the same volume of approximately 2M sodium hydroxide solution. Note the colour of the precipitate of mercury(II) oxide and compare it with that of mercury(II) oxide taken from a bottle. The difference in colour is simply due to a difference in particle size. Note that mercury(II) hydroxide is not formed since it is unstable with respect to decomposition into mercury(II) oxide and water.

Is mercury(II) oxide soluble in an excess of sodium hydroxide solution? How would you classify mercury(II) oxide?

(b) Action of heat on mercury(II) oxide

Place about $0.1 g$ of mercury(II) oxide in an ignition tube and heat gently **in a fume cupboard.** Test for oxygen with a glowing splint and note the formation of silvery globules of mercury on the cooler regions of the test-tube.

13.13 Mercury(II) Sulphide, HgS

Formation of mercury(II) sulphide

Place about $3 cm^3$ of a solution of mercury(II) chloride in a test-tube and pass into it a stream of hydrogen sulphide. Note the colour of the mercury(II) sulphide precipitate and find out whether it is soluble in hydrochloric acid.

13.14 Mercury (II) Chloride, $HgCl_2$

Reduction of mercury(II) chloride with tin(II) chloride

Place about 2 cm^3 of a solution of mercury(II) chloride in a test-tube and add to it a solution of tin(II) chloride. Note the colour of the precipitate. Finally add tin(II) chloride in excess and warm if necessary.

Mercury(II) ions are reduced in two stages by the tin(II) ions; what are the intermediate and final precipitates? Are your results consistent with the following standard redox potentials?

$$Sn^{4+} + 2e^- \rightarrow Sn^{2+} \qquad E^\ominus = +0{\cdot}14 \text{ V}$$
$$2Hg^{2+} + 2e^- \rightarrow Hg_2^{2+} \qquad E^\ominus = +0{\cdot}92 \text{ V}$$
$$Hg_2^{2+} + 2e^- \rightarrow 2Hg \qquad E^\ominus = +0{\cdot}80 \text{ V}$$

13.15 Mercury(II) Iodide, HgI_2

(a) Formation of mercury(II) iodide and its solubility in excess potassium iodide solution

Place 25 drops of 0·2M mercury(II) chloride solution in a test-tube and add, drop by drop, using the same clean dropping tube, some M potassium iodide solution. Note the colour of the mercury(II) iodide precipitate. Go on adding the potassium iodide solution, drop by drop with shaking, until the precipitate just dissolves. From your results calculate the value of n (to the nearest whole number) in the following equation and hence determine the formula of the complex ion:

$$Hg^{2+} + nI^- \rightleftharpoons [HgI_n]^{(n-2)-}$$

To this clear solution add approximately 2M sodium hydroxide solution, drop by drop, and note whether any precipitate of mercury(II) oxide is formed (see Experiment 13.12(a), p. 195). How do you explain the result?

(b) The action of heat on mercury(II) iodide

Rub some solid mercury(II) iodide on a piece of cardboard and then heat it by holding it close to a Bunsen burner. Note the colour change that occurs, and also the colour of the mercury(II) iodide after it has cooled. Now scratch the mercury(II) iodide with a piece of wire and note the result. Can you explain this phenomenon?

MERCURY(I) COMPOUNDS

An interesting feature of mercury is its ability to form the Hg_2^{2+} ion in which two mercury atoms are united by a single covalent bond, i.e. the simple Hg^+ ion does not exist. One piece of evidence in favour of the Hg_2^{2+} ion as opposed to the Hg^+ ion is the fact that mercury(I) compounds are not paramagnetic, but if the formula of the mercury(I) ion was Hg^+

this would contain an unpaired $6s$ electron. In Hg_2^{2+} both $6s$ electrons originating from each mercury atom are paired and this explains its lack of paramagnetism.

Attempts to prepare mercury(I) hydroxide, mercury(I) oxide and mercury(I) sulphide have failed and, in order to understand the reason for this, it is necessary to consider the equilibrium:

$$Hg_2^{2+} \rightleftharpoons Hg + Hg^{2+}$$

The above disproportionation reaction has an equilibrium constant K of about 6.0×10^{-3} at 298K, i.e.

$$K = \frac{[Hg^{2+}]}{[Hg_2^{2+}]} = 6.0 \times 10^{-3}$$

The low value for the equilibrium constant implies that under normal conditions there is little tendency for the mercury(I) ion to disproportionate into the mercury(II) ion and mercury. However, the addition of an anion, for example the sulphide ion, which forms a more insoluble compound with the mercury(II) ion than with the mercury(I) ion, causes the disproportionation reaction to proceed further to the right:

$$Hg_2^{2+} \rightleftharpoons Hg + Hg^{2+}$$
$$+$$
$$S^{2-}$$
$$\updownarrow$$
$$HgS$$

The addition of hydrogen sulphide to an aqueous solution of a mercury(I) salt thus results in the precipitation of mercury(II) sulphide and free mercury:

$$Hg_2^{2+} + H_2S \rightarrow Hg + HgS + 2H^+$$

Mercury(I) compounds form few complexes principally because mercury(II) compounds form very stable ones with many ligands; thus once again the disproportionation reaction is driven well over to the right:

$$Hg_2^{2+} \rightleftharpoons Hg + Hg^{2+}$$
$$+$$
$$\text{Complexing ligands}$$
$$\updownarrow$$
$$Hg^{2+} \text{ complex}$$

Two of the best known compounds of mercury(I) are mercury(I) chloride and mercury(I) nitrate.

13.16 Mercury(I) Chloride, Hg_2Cl_2

(a) Formation of mercury(I) chloride

Place about $2 \, cm^3$ of a solution of mercury(I) nitrate in a test-tube and add to it about the same volume of approximately 2M hydrochloric acid. Note the result.

(b) Action of aqueous ammonia solution on mercury(I) chloride

To the precipitate of mercury(I) chloride obtained above, add some approximately 2M aqueous ammonia and note the formation of a black deposit of mercury and a compound of mercury(II), $HgNH_2Cl$, which contains the mercury–nitrogen covalent bond.

$$Hg_2Cl_2 + 2NH_3 \rightarrow Hg + HgNH_2Cl + NH_4Cl$$

13.17 Mercury(I) Nitrate, $Hg_2(NO_3)_2$

Reduction of mercury(I) nitrate with tin(II) chloride solution

Make a solution of mercury(I) nitrate in water, adding some approximately 2M nitric acid if necessary to obtain a clear solution. To about 2 cm³ of this solution add an excess of tin(II) chloride solution and warm if necessary. Note the result, and then write a balanced ionic equation for the reaction between Hg_2^{2+} ions and Sn^{2+} ions in aqueous solution.

13.18 Summary

(a) Zinc occurs as the carbonate and the sulphide and these are first converted into zinc oxide which is then reduced to zinc by carbon. Cadmium is obtained during the purification of the zinc by distillation.

(b) Zinc and cadmium are fairly reactive metals and will displace hydrogen from dilute hydrochloric acid. The pure metals, however, resist attack by the dilute acid, since they exhibit a high overpotential to the discharge of hydrogen (an activation energy effect).

(c) The oxides may be obtained by heating either the carbonate or the nitrate. Zinc oxide is white (yellow when hot) and cadmium oxide is brown. Both oxides readily dissolve in dilute acids, but zinc oxide, in addition, will dissolve in sodium hydroxide solution, i.e. zinc oxide is amphoteric while cadmium oxide is basic.

(d) The hydroxides may be obtained as white solids by the addition of sodium hydroxide solution to a solution of their salts. Zinc hydroxide, like the oxide, is amphoteric, while cadmium hydroxide is basic. Both hydroxides dissolve in aqueous ammonia forming complex ions of the general formula $[M(NH_3)_4]^{2+}$.

(e) The sulphides may be precipitated by passing hydrogen sulphide through solutions of their salts. Zinc sulphide which is white is readily soluble in dilute acid, whereas cadmium sulphide which is yellow is only soluble in more concentrated acid solutions.

(f) Both metals form deliquescent nitrates which readily decompose on heating into the metallic oxide, nitrogen dioxide and oxygen.

(g) Mercury occurs as the sulphide from which the metal is extracted by heating in air.

(h) Mercury, unlike zinc and cadmium, is not attacked by dilute non-oxidising acids. It is attacked by nitric acid; excess of concentrated nitric acid tends to give nitrogen dioxide and mercury(II) nitrate, but with dilution, and in the presence of excess mercury, nitrogen oxide and mercury(I) nitrate tend to predominate.

(i) Mercury exhibits oxidation states of $+2$ and $+1$; in the latter, two mercury atoms are united by a single covalent bond as in the $[Hg-Hg]^{2+}$ ion.

(j) Mercury(II) oxide may be precipitated by adding sodium hydroxide solution to a solution of mercury(II) chloride. It does not dissolve in an excess of sodium hydroxide solution and is thus exclusively basic. Mercury(II) hydroxide does not exist.

(k) Mercury(II) sulphide may be precipitated as a black solid by passing hydrogen sulphide through a solution of mercury(II) chloride. It is not soluble in hydrochloric acid, unlike zinc sulphide and cadmium sulphide.

(l) Mercury(II) chloride solution is an oxidising agent; for example, tin(II) chloride reduces it to mercury(I) chloride and then to mercury.

(m) Mercury(II) iodide may be precipitated by adding potassium iodide solution to a solution of mercury(II) chloride. It is a red solid which dissolves in an excess of potassium iodide solution, forming the complex ion $[HgI_n]^{(n-2)-}$. The action of heat on mercury(II) iodide converts it into a yellow polymorphic form. On cooling, the change is reversed slowly, but it may be accelerated by scratching the solid.

(n) The best known compounds of mercury(I) are the chloride, Hg_2Cl_2, which is very insoluble in water, and the nitrate, $Hg_2(NO_3)_2$, which is soluble in water. Both compounds are readily reduced to mercury by the addition of tin(II) chloride solution.

APPENDIX I
CHEMICALS REQUIRED

Acetylacetone
Aluminium (filings, foil, powder)
Aluminium chloride (anhydrous)
Aluminium nitrate
Aluminium oxide (chromatographic grade)
Aluminium sulphate
Ammonium chloride
Ammonium sulphate
Antimony(III) oxide
Antimony(V) oxide
Antimony trichloride
Arsenic(III) oxide
Arsenic(V) oxide
Barium carbonate
Barium chloride
Barium hydroxide
Barium nitrate
Barium oxide
Barium peroxide
Benzene
Bismuth (granulated)
Bismuth carbonate
Bismuth nitrate
Bismuth(III) oxide
Bismuth oxychloride
Bleaching powder
Boron (amorphous)
Boron(III) oxide
Bromine (liquid)
Caesium chloride
Cadmium (granulated)
Cadmium carbonate
Cadmium nitrate
Cadmium sulphate
Calcium (pellets)
Calcium carbide
Calcium carbonate
Calcium chloride
Calcium ethanoate (acetate)
Calcium hydride
Calcium hydroxide
Calcium nitrate
Calcium oxide
Calcium phosphide
Carbon (wood charcoal powder, activated charcoal)
Carbon disulphide
Chlorosulphonic acid
Chromium (lumps)
Chromium(III) sulphate
Cobalt(II) chloride
Cobalt(II) nitrate
Cobalt(II) sulphate

Copper (clippings, foil, powder)
Copper(II) carbonate
Copper(II) chloride
Copper(II) methanoate (formate)
Copper(II) nitrate
Copper(I) oxide
Copper(II) oxide
Copper(II) sulphate
1,2-diaminoethane
Dichlorodimethylsilane
Diethyl ether
Dilead(II) lead(IV) oxide (red lead)
Disodium hydrogen orthophosphate
Disodium hydrogen orthophosphite
Ethanoic acid (acetic acid)
Ethanoic anhydride (acetic anhydride)
Ethanol
Ethylenediaminetetra-acetic acid (EDTA)—
    the disodium salt
Glucose
Hydrazine hydrate
Hydrazine sulphate
2-Hydroxybenzoic acid (salicylic acid)
Iodine
Iodine(V) oxide
Iron (filings)
Iron(II) disulphide (iron pyrites)
Iron(III) oxide
Iron(II) sulphate
Iron(II) sulphide
Lead (granulated)
Lead(II) carbonate
Lead(II) nitrate
Lead(II) oxide
Lead(IV) oxide
Lithium
Lithium carbonate
Lithium chloride
Lithium hydride
Lithium hydroxide
Lithium nitrate
Magnesium (ribbon, turnings)
Magnesium carbonate
Magnesium chloride
Magnesium ethanoate (acetate)
Magnesium hydroxide
Magnesium nitrate
Magnesium oxide
Magnesium sulphate
Manganese (lumps)
Manganese(II) chloride
Manganese(IV) oxide
Manganese(II) sulphate
Mercury(II) iodide
Mercury(I) nitrate
Mercury(II) oxide
Methanoic acid (formic acid)
Methanol
Nickel (foil)

Nickel sulphate
Orthoboric acid
Orthophosphoric acid
Orthophosphorous acid
Phosphorus (red, white)
Phosphorus(V) oxide
Phosphorus pentachloride
Phosphorus trichloride
Potassium antimonate
Potassium bromate(V)
Potassium bromide
Potassium carbonate (Analar)
Potassium chlorate(V)
Potassium chloride
Potassium dichromate
Potassium hydrogen carbonate
Potassium hydroxide
Potassium iodate(V)
Potassium iodide
Potassium manganate(VI)
Potassium nitrate
Potassium permanganate
Potassium persulphate
Potassium sulphate
Rubidium chloride
Silicon (amorphous)
Silicon(IV) oxide
Silicon tetrachloride
Silicone fluid (MS 1107, marketed by Hopkins & Williams Ltd.)
Soda lime
Sodium
Sodium arsenate
Sodium arsenite
Sodium bismuthate
Sodium bromide
Sodium carbonate (Analar)
Sodium chloride
Sodium dihydrogen orthophosphate
Sodium fluoride
Sodium hydrogen carbonate
Sodium hydroxide
Sodium iodide
Sodium methanoate (formate)
Sodium nitrate
Sodium nitrite
Sodium oxalate
Sodium potassium tartrate
Sodium pyrosulphite
Sodium sulphate
Sodium sulphide
Sodium sulphite
Sodium tetraborate (borax)
Strontium carbonate
Strontium chloride
Strontium hydroxide
Strontium nitrate
Strontium oxide
Sulphur

Sulphur dichloride oxide (thionyl chloride)
Tetrachloromethane
Tin (foil, granulated)
Tin(II) chloride
Tin(IV) chloride (hydrated)
Tin(II) oxalate
Turpentine
Vanadium(V) oxide
Water glass
Zinc (foil, granulated)
Zinc carbonate
Zinc nitrate
Zinc oxide
Zinc sulphate

APPENDIX II

SOLUTIONS REQUIRED (I)

The concentrations of solutions given in this list need only be very approximate.

Name	Formula	Relative formula mass	Concentration
Alizarin			0·1 g in 100 cm^3 ethanol
Aluminium sulphate	$Al_2(SO_4)_3$	630	0·05M
*Ammonium metavanadate	NH_4VO_3	117	0·1M
Ammonium molybdate		1236	0·1M
Ammonium thiocyanate	NH_4CNS	76	0·1M
Ammonia (880)			Saturated solution
Ammonia (aqueous)			2M
Barium chloride	$BaCl_2 \cdot 2H_2O$	244	0·1M, 0·2M, 0·4M, 0·5M
Barium hydroxide	$Ba(OH)_2 \cdot 8H_2O$	316	0·05M
Bromine water			Saturated solution
Calcium chloride	$CaCl_2$	111	0·1M, 0·2M, 0·4M, 0·5M
Chlorine water			Saturated solution
Chromium(III) potassium sulphate	$KCr(SO_4)_2 \cdot 12H_2O$	500	0·05M
Cobalt(II) chloride	$CoCl_2 \cdot 6H_2O$	238	0·1M
Copper(II) sulphate	$CuSO_4 \cdot 5H_2O$	250	0·1M, 0·3M
Disodium hydrogen orthophosphate	$Na_2HPO_4 \cdot 12H_2O$	358	0·2M
Hydrochloric acid			1M, 2M, Concentrated
Hydrogen peroxide			20 volume
Iodine in potassium iodide solution			0·1M with respect to both I_2 and KI
Iron(II) sulphate	$FeSO_4 \cdot 7H_2O$	278	0·1M

*Dissolve the solid in water and add a little, approximately 0·5M, sulphuric acid. This gives a red precipitate which should be dissolved in the minimum volume of approximately 2M sodium hydroxide solution.

Appendix II (Contd.)

Name	Formula	Relative formula mass	Concentration
Iron(III) chloride	$FeCl_3 \cdot 6H_2O$	270	0·1M
Iron(III) nitrate	$Fe(NO_3)_3 \cdot 9H_2O$	404	0·5M
Lead(II) nitrate	$Pb(NO_3)_2$	331	0·1M, 0·2M, 0·5M
Lime water			Saturated solution
Lithium chloride	LiCl	43	0·2M
Magnesium chloride	$MgCl_2 \cdot 6H_2O$	203	0·2M, 0·4M, 0·5M
Magnesium sulphate	$MgSO_4 \cdot 7H_2O$	247	0·1M
Manganese(II) sulphate	$MnSO_4 \cdot 4H_2O$	223	0·1M
Mercury(II) chloride	$HgCl_2$	272	0·1M
Nickel sulphate	$NiSO_4 \cdot 7H_2O$	281	0·1M
Nitric acid			2M, Concentrated
Potassium chloride	KCl	75	0·1M
Potassium chromate	K_2CrO_4	194	0·2M
Potassium dichromate		294	0·1M
Potassium iodide		166	0·5M, 2M
Potassium permanganate	$KMnO_4$	158	0·02M
Silver nitrate	$AgNO_3$	170	0·1M
Sodium carbonate	Na_2CO_3	106	0·2M, 1M
Sodium fluoride	NaF	42	0·2M
Sodium halate(I) (hypochlorite)			2M
Sodium hydrogen carbonate	$NaHCO_3$	84	1M
Sodium hydroxide	NaOH	40	2M
Sodium potassium tartrate	$KNaC_4H_4O_8 \cdot 4H_2O$	282	0·4M
Sodium potassium tartrate (alkaline)	$KNaC_4H_4O_8 \cdot 4H_2O$	282	1·3M with respect to sodium potassium tartrate and 4M with respect to NaOH
Sodium thiosulphate	$Na_2S_2O_3 \cdot 5H_2O$	248	0·2M
Strontium chloride	$SrCl_2 \cdot 6H_2O$	266	0·2M, 0·5M
Sulphuric acid			0·2M, 0·4M, 1M, Concentrated
Zinc sulphate	$ZnSO_4 \cdot 7H_2O$	288	0·1M

APPENDIX III

SOLUTIONS REQUIRED (II)

The concentrations of solutions given in this list must be accurately known but need not be exactly those given in the fourth column.

Name	Formula	Relative formula mass	Concentration
Ammonium iron(II) sulphate	$(NH_4)_2Fe(SO_4)_2 \cdot 6H_2O$	392	0·1M
Copper(II) sulphate	$CuSO_4 \cdot 5H_2O$	250	0·1M, 1M
EDTA (disodium salt)		372	0·05M
Hydrochloric acid			1M
Iodine (in aqueous potassium iodide)	I_2	254	0·05M
Manganese(II) chloride	$MnCl_2 \cdot 4H_2O$	198	0·16M
Mercury(II) chloride	$HgCl_2$	272	0·2M
Nickel sulphate	$NiSO_4 \cdot 7H_2O$	281	0·05M
Potassium chloride	KCl	74·5	0·1M
Potassium dichromate	$K_2Cr_2O_7$	294	0·018M
Potassium dichromate/ potassium permanganate mixture	$K_2Cr_2O_7/KMnO_4$		0·02M with respect to both salts
Potassium iodate(V)	KIO_3	214	0·0167M
Potassium iodide	KI	166	0·2M, 1M
Potassium permanganate	$KMnO_4$	158	0·018M, 0·12M
Silver nitrate	$AgNO_3$	170	0·1M
Sodium ethanoate (acetate)	$CH_3COONa \cdot 3H_2O$	136	2M
Sodium thiosulphate	$Na_2S_2O_3 \cdot 5H_2O$	248	0·01M, 0·1M
Zinc sulphate	$ZnSO_4 \cdot 7H_2O$	288	1M

Index

Aluminium, 40, 42–43
Aluminium chloride, 47–48
Aluminium hydroxide, 46–47
Aluminium iodide, 42
Aluminium oxide, 43–46
Aluminium potassium sulphate, 49
Ammonia, 75, 76
Ammonium chloride, 75–76
Ammonium hexachloroplumbate(IV), 70
Antimony, 74, 87
Antimony(III) oxide, 88
Antimony(V) oxide, 88–89
Antimony pentachloride, 92
Antimony(III) sulphide, 90–91
Antimony(V) sulphide, 90–91
Antimony trichloride, 92
Arsenates, 89–90
Arsenic, 74, 87
Arsenic(III) oxide, 88
Arsenic(V) oxide, 88–89
Arsenic(III) sulphide, 90–91
Arsenic(V) sulphide, 91
Arsenic trichloride, 92
Arsenites, 89

Barium, 31
Barium carbonate, 36
Barium chloride (hydrated), 35–36
Barium chromate, 36
Barium ferrate(VI), 169–170
Barium hydroxide, 33, 34
Barium oxide, 33
Barium peroxide, 33
Barium phosphate, 36
Barium sulphate, 36
Barium d-tartrate, 172
Bis(dimethylglyoximato)nickel(II), 177
Bismuth, 74, 87
Bismuth carbonate (basic salt), 93
Bismuth hydroxide, 90
Bismuth nitrate, 93
Bismuth(III) oxide, 88
Bismuth(III) sulphide, 91–92
Bismuth trichloride, 92
Borate glasses, 41
Boric acid (see orthoboric acid)
Boron, 40
Boron(III) oxide, 40–41
Bromine,
    action of alkalis and acids on aqueous solution of, 136
    formation of, 134
    oxidising action of, 137–138
    reactions of, 135–136
    solubility of, 134, 135

Cadmium, 191–192
Cadmium carbonate, 194
Cadmium hydroxide, 193
Cadmium nitrate, 194
Cadmium oxide, 192
Cadmium sulphide, 193
Caesium, 20
Calcium, 31–33
Calcium carbonate, 35
Calcium chloride (hydrated), 35–36
Calcium chromate, 36
Calcium ethanoate (acetate), 37
Calcium hydride, 34
Calcium hydroxide, 33, 34
Calcium nitrate, 35
Calcium nitride, 34–35
Calcium oxide, 33
Calcium phosphate, 36
Calcium sulphate, 36–37
Carbon, 52–53
Carbon dioxide, 54, 56, 57
Carbon monoxide, 55
Carbonates, 54
Carbonic acid, 56
Chlorides, 15–16, 142
Chlorine,
    action of acids and alkalis on aqueous solutions of, 136–137
    formation of, 133–134
    oxidising action of, 137–138
    reactions of, 135–136
    solubility of, 134, 135
Chlorosulphonic acid, 126
Chromium,
    chromium(II) compounds, 165
    chromium(III) compounds, 164–165
    chromium(VI) compounds, 163–164
    formation of chromium(VI) dichloride dioxide, 164
Cobalt,
    cobalt(II) compounds, 173
    cobalt(III) compounds, 170–173
Copper,
    copper(I) compounds, 183–185
    copper(II) compounds, 180–183
    extraction of, 179
    reactions of, 180
Copper(II) carbonate, 182

Copper(I) chloride, 184
Copper(II) hydroxide, 181
Copper(I) iodide, 185
Copper(II) methanoate (formate), 56–57
Copper(II) oxide, 181–182
Copper(II) salts, 182–183
Copper(II) sulphide, 182

Dichlorobis(1, 2-diaminoethane) cobalt(III) chloride, 170–171
Dilead(II) lead(IV) oxide, 64, 65, 66
Disodium hydrogen orthophosphate, 84
Disodium hydrogen orthophosphite, 84
Disulphur dichloride, 113–114, 129

Electrode potential, 99–105
Elements,
  chemical properties of, 10–13
  physical properties of, 7–10

Hydrazine hydrate, 76
Hydrazine sulphate, 77
Hydrides,
  action of water on, 14–15
  physical nature of, 15–16
Hydrogen bromide, 138–141
Hydrogen carbonates, 54
Hydrogen chloride, 138–141
Hydrogen iodide, 138–141
Hydrogen peroxide, 111–112
Hydrogen sulphide, 115–117

Iodine,
  action of acids and alkalis on aqueous solution of, 136–137
  extraction of, 145
  formation of, 134
  oxidising action of, 137–138
  reactions of, 135–136
  solubility of, 134–135
Iodine monochloride, 146
Iodine(V) oxide, 142
Iodine trichloride, 146
Iron,
  iron(II) compounds, 169–170
  iron(III) compounds, 169–170
  iron(VI) compounds, 169–170
Iron(II) disulphide, 119

Lead, 52, 64
Lead(II) chloride, 69–70
Lead(II) chromate, 70–71
Lead(IV) ethanoate (acetate), 71
Lead(II) hydroxide, 66
Lead(IV) oxide, 65–66
Lead(II) sulphate, 71
Lead(II) sulphide, 67
Lithium, 20–23

Lithium carbonate, 26–27
Lithium chloride, 28
Lithium fluoride, 28
Lithium hydride, 26
Lithium hydroxide, 24
Lithium monoxide, 23–24
Lithium nitrate, 27

Magnesium, 31–33
Magnesium carbonate, 35–36
Magnesium chloride (hydrated), 35–36
Magnesium chromate, 36
Magnesium ethanoate (acetate), 37
Magnesium hydroxide, 33–34
Magnesium nitrate, 35
Magnesium nitride, 34–35
Magnesium oxide, 31–32, 33
Magnesium phosphate, 36
Magnesium sulphate, 36–37
Manganese,
  manganese(II) compounds, 169
  manganese(III) compounds, 168–169
  manganese(IV) compounds, 167
  manganese(VI) compounds, 166–167
  manganese(VII) compounds, 165–166
Manganese(IV) oxide, 167
Manganese(II) sulphate,
  paramagnetism of, 160–161
Mercury, 194–195
Mercury(I) chloride, 197–198
Mercury(II) chloride, 196
Mercury(II) iodide, 196
Mercury(I) nitrate, 198
Mercury(II) oxide, 195
Mercury(II) sulphide, 195
Methane, 62–63
Methyl orthoborate, 41

Nickel,
  nickel(II) compounds, 173
  nickel(III) compounds, 173
Nickel(II)–EDTA complex ion, 174–175
Nitrates, 79–80
Nitric acid, 79
Nitrites, 78–79
Nitrogen, 74–75
Nitrous acid, 78

Orthoboric acid, 41
Oxidation, 96–105
Oxides,
  physical nature of, 15–16
Oxygen, 107–108

Peroxodisulphuric acid, 128
Phosphates, 84–86
Phosphine, 82
Phosphites, 83–84

Phosphoric acid, 84
Phosphorous acids, 83
Phosphorus, 74, 80–81
Phosphorus(III) oxide, 82–83
Phosphorus(V) oxide, 82–83
Phosphorus pentachloride, 86–87
Potassium, 20–21
Potassium bromate(V), 143–144
Potassium carbonate, 26–27
Potassium chromate, 163–164
Potassium chromium(III) sulphate, 164–165
Potassium chlorate(V), 143–144
Potassium chloride, 28
Potassium dichromate, 163, 164–165
Potassium hydrogen carbonate, 27
Potassium hydroxide, 24
Potassium iodate(V), 143–145
Potassium manganate(VI), 166–167
Potassium nitrate, 27
Potassium permanganate, 165–166
Potassium peroxodisulphate (persulphate), 128–129

Redox potentials, 99–105
Redox reactions, 98–105
Reduction, 96–105
Rubidium, 20

Salt hydrolysis, 109–110
Silanes, 63
Silica gel, 58
Silicates, 59–60
Silicic acid, 58
Silicon, 52–53
Silicon(IV) oxide, 54, 56
Silicon tetrachloride, 63
Silicones, 61–62
Silver, 185–186
Silver bromide, 187
Silver chloride, 187
Silver chromate, 188
Silver iodide, 187
Silver oxide, 186
Silver sulphide, 187–188
Sodium, 20–23
Sodium arsenate, 89–90
Sodium arsenite, 89
Sodium bismuthate, 90
Sodium carbonate, 26–27, 57
Sodium chlorate(I), 143
Sodium chloride, 28
Sodium dihydrogen orthophosphate, 84
Sodium hydrogen carbonate, 27, 57
Sodium hydroxide, 24
Sodium metaphosphate, 86
Sodium methanoate (formate), 56
Sodium nitrate, 27
Sodium peroxide, 23–24

Sodium sulphide, 118
Sodium thiosulphate, 127–128
Strontium, 31
Strontium carbonate, 35, 36
Strontium chloride (hydrated), 35–36
Strontium chromate, 36
Strontium hydroxide, 33–34
Strontium nitrate, 35
Strontium oxide, 33
Strontium phosphate, 36
Sulphates, 125
Sulphides, 118
Sulphites, 120–122
Sulphur, 107, 113–115
Sulphur dichloride dioxide (sulphuryl chloride), 126
Sulphur dichloride oxide (thionyl chloride), 113–114, 129–130
Sulphur dioxide, 119–120
Sulphur trioxide, 120
Sulphuric acid, 122–124
Sulphurous acid, 120–121

Tetrachloromethane, 63, 109
Tetrasodium pyrophosphate, 85
Tin, 52, 64
Tin(II) chloride, 69
Tin(IV) chloride, 67–69
Tin(II) hydroxide, 66
Tin(II) oxalate, 65
Tin(II) oxide, 64–65
Tin(IV) oxide, 64–65
Tin(II) sulphide, 67
Tin(IV) sulphide, 67
Transition elements,
    catalytic activity of, 158–160
    colours of their salts, 151–153
    complex ion formation, 153–157
    extraction of, 150
    paramagnetism of their compounds, 160
    reactions of them and their salts, 150–151
    variable oxidation states of, 157–158
Tris(acetylacetonato) manganese(III), 168–169
Tris(1,2-diaminoethane) cobalt(III) chloride d-tartrate, 172
Tris(1,2-diaminoethane) cobalt(III) iodide monohydrate, 171–172

Vanadium, 162

Water, 108–111

Zinc, 191–192
Zinc carbonate, 194
Zinc hydroxide, 193
Zinc nitrate, 194
Zinc oxide, 192
Zinc sulphide, 193